I0033369

Humphrey Ridley

The Anatomy of the Brain

Containing its mechanism and physiology, together with some new discoveries and

corrections of ancient and modern authors upon that subject

Humphrey Ridley

The Anatomy of the Brain

Containing its mechanism and physiology, together with some new discoveries and corrections of ancient and modern authors upon that subject

ISBN/EAN: 9783337240882

Printed in Europe, USA, Canada, Australia, Japan

Cover: Foto ©berggeist007 / pixelio.de

More available books at **www.hansebooks.com**

Tractatum hunc cui titulus

The Anatomy of the BRAIN;

Dignum Judicamus qui Imprimatur,

Thomas Burwell, Præfes.

Dat. ex Ædibus
Collegii in Co-
mitiis Cenfor.
Sept. 7. 1694

Samuel Collins,
Fred. Slare,
William Dawes,
Tancred Robinfon

} Cenfores.

THE
ANATOMY
OF THE
BRAIN.

Containing its
Mechanism and *Physiology* ;
Together with some
New Discoveries and Corrections
OF
Ancient and Modern Authors .
Upon that SUBJECT.

To which is annex'd a particular Account of
ANIMAL FUNCTIONS
AND
Muscular Motion.

*The Whole illustrated with Elegant Sculptures
after the life.*

By *H. RIDLEY*, Coll. Med. Lond. Soc.

LONDON:
Printed for *Sam. Smith* and *Benj. Walfo.d*,
Printers to the Royal Society, at the *Princes
Arms* in St. *Paul's* Church-yard, 1695.

Spectatissimo Doctissimoque iro

D. D. JOHANNI LAWSON

Collegii Regalis Medicorum London.

Presidi Dignissimo

SOCIIS,

Et inter eos speciatim

CENSORIBUS

Vel eo nomine Clarissimis

SAMUELI COLLINS,
RICHARDO TORLESS,
EDVARDO TYSON,
MARTINO LISTER.

NECNON

D. D. *Electoribus Meritissimis*

Omnibus & Singulis.

Tam præ Universali Exquisita sua eruditione,
quam Artis Apollineæ *Praxi fœlicissima*
longe Celeberrimis

CÆTERIS *Denique,*

Egregiis Viris
Inclytissimæ hujus Societatis ascriptis

Paginas has eorum jussu in lucem prodeuntes,
Honoris & Obsequii Ergo
quam Humillimè Offert,

H· R.

THE
PREFACE
TO THE
Reader.

THAT Reaſon which, upon firſt thoughts, ſeemed of moſt force to diſſwade me from engaging my ſelf upon the Subject I have made choice of in theſe few following Sheets, (which was, its having been already undertaken by two ſo eminent Perſons, as the late *Willis*, and the preſent *Vieuſſenius*) upon ſecond became the greateſt motives to it. Seeing that even after the beſt Proofs they have either of them been able to give of *Skill*

or

or *Industry* upon this Subject, there hath yet escap'd undisco- ver'd both a great deal of the Materials which Nature is wont to furnish for the framing of Parts, and Contrivance too in ranging of them, in order to bring about that great design of making them all contribute their share to the conservation of the whole.

The truth of this becom- ing still more evident whilst I became more conversant in *Dis- section*, after some time, put me upon an endeavour, by a deeper Scrutiny, to discover something more than what as yet had come to light : and this I undertook so much the more vigorously, as by how much I reckon'd it more pre- ferrable to contribute my Mite towards the perfecting of a Work already so happily be- gun

gun and fuccefsfully carried on,
than to break the Ice only
(the common Fate of the firft
attempt) of another.   With
what fuccefs I have done it
the Reader muft be Judge.

Through the whole defcri-
ption of Parts I have offer'd
nothing but Matter of Fact,
and have taken all poffible
care to avoid being impos'd
upon my felf, by making Experiments in proportion to my
Doubts.   Some of them have
been upon Subjects in their
natural, fome in their morbid
eftate, fome upon thofe of Un-
timely Death ; and on thofe laft
fometimes whilft the natural
Fluids remained in their proper
Veffels, though after a preternatural manner occafion'd by
Strangulation ; fometimes when
in the room thereof, other
Bodies have been introduc'd
by

by Injection, as *Tinged Wax* and *Mercury*, the first of which by its confistence chiefly, the other by its permanent nature and colour, contribute mightily towards bringing to view the most minute ramifications of Veffels, and fecreteft receffes of Nature,

By this various difpofition of the Subject it is that fo great Difficulties are overcome in fearch after Truth, many things appearing oftentimes very plain in one ftate, which either lay concealed, or feemed otherwife modified in any of the other.

The Figures were delineated by the hand of that Compleat Anatomift Mr. *Cowper* the Surgeon, whofe great Skill in Diffection renders that Talent fo fortunate both to himfelf and his Friends : and how exactly
that

that Work is performed, I fub-
mit to the Severeſt Cenſure of
any who will be at the pains
to compare any of the Cuts to
the life.

What I have ſaid upon the
Phyſiologia, in relation to *Nutri-*
tion and *Muſcular Motion*, depends on Microſcopical Obſer-
vations ; and as to the *Poſtula-*
tum on which they both depend, though at firſt ſight it
may appear ſurprizing, yet I
am confident it will become
far leſs ſo to thoſe who have
been acquainted with what
hath been ſaid of the Vaſcular
Compages of . Plants ·by *Mal-*
pighius and *Grew*, and of ſeveral other Subjects by *Lewen-*
hoeck.

And to conclude, I muſt
confeſs I have been the better
ſatisfied with it my ſelf, ſince I
met with ſome Paſſages in the
Works

Works of thofe learned Micro-
graphifts **Dr.** *Power*, **and of**
Mr. *Hooke*, relating to this Subject, in which laft, the *medium*
made ufe of for folution of that
famous *Phænomenon* of that Plants
contraction at the firft appulfe
of **Touch** from external Objects, as well as the manner of
its acting, is the fame with that
made ufe of here as a *Poftula-*
tum, upon which the whole of
what is faid about *Mufcular Mo-*
tion is built : Altho' at the fame
time I am fenfible 'tis not fo apply'd in that place by the aforefaid Author, whofe opinion in
reference to *Mufcular Motion*
(being the fame with that of
Dr. *Mayow* already taken notice
of in the following Sheets) is ex-
prefly otherwife in the account
he gives of thofe natural *Hygro-*
meters the Beards of *Wild Oats*,
of all the forts of *Cranes Bills*
and

and *Cats Guts*, conformable to the manner of Nature's acting on which, in order to make them proper *Indexes* of the variousChanges of Weather (*viz.* by *wreathing* and *unwreathing*) he suppofes that to be of *Mufcular Motion*.

I have quoted Authors, not out of oftentation, but both for their Truth and Errors, to the end that at the fame time we may fee it reafonable and convenient to read all they fay, we may be render'd cautious how we believe; and to put us in mind, that as we find fomething done to our hands by thofe who have gone before, there is reafon we fhould do fomething for thofe who are to come after.

THE

THE

INTRODUCTION.

Howfoever the Controverfie may ftand amongft Learned Men, about the Method and Order which Nature makes ufe of in the framing the different Parts of Animals, efpecially as to precedency of Time, fome of them fuppofing a rudimentary delineation, or pre-exiftence of the whole, which, as the Ingenious Bruner hath rightly obferved, muft neceffarily imply an actual exiftence of the whole Race of Mankind at once, either in the Tefticle or Ovarium of Eve, according to the Learned Harvey, Malpighius, Swammardam, &c. or in that of Adam, according to Lewenhoeck, Dr. Garden, and feveral others, and confequently muft needs alfo infer an extinction of the fame Progeny, as foon as the number of thofe humane Germens or Animalcles fhall be exhaufted; others a gradual formation of parts,

one

one after another, by an intestine motion begun and carried on from the
time of coition, by the subtle matter
in the Cicatricula of the Egg : I see
no reason to make my self a Party on
either side at this time, seeing the
fineness of structure and dignity of
functions are sufficient to give prefe-
rence to one above another, and to render
it more worthy of a particular consideration. And this part I take to be
the Brain, the delicacy of whose Structure is such, that with no little resem-
blance to its divine Author, whilst it
gives us the greatest and clearest discoveries of other things, lies most concealed it self.

* And seeing all that Mystick Knowledge, which in ancient times, in the
eyes especially of the Vulgar, appeared
meer Necromancy or Witchcraft, as
well as all the Curious Discoveries of
more modern Ages upon the whole subject of Nature, now going under the
more familiar and proper term of Refined Sence, or Philosophy, hath been
meerly owing to a more acurate knowledge of the parts and modification of
Matter, I see not any more likely
way of conquering the difficulties yet
be-

behind upon any particular *subject*,
than the endeavouring after a further and more nice *scrutiny* into it
by *such* means and experiments as
serve to bring its most minute parts
and texture under the test of Sence,
which so *assisted*, doth the same office
to the discerning faculty as good arti-
ficial Glasses do to it, bringing the
Object and Judgment to such a near-
ness, that even the first Link of the
Chain becomes discernable, and the
mechanical proceedings of Nature so
highly instructive to the Understand-
ing, in its finding out and assigning
proper Causes to Effects much more
obvious and intelligible.

I shall therefore treat this Noble
Part after the aforesaid manner, with
all the Justice I can, leaving those
invisible, and almost divine things
called Animal Spirits, to be treated of more at large, by those more
illuminated Philosophers, who see best
when their Eyes are shut, and content
my self with making an inquiry into,
and giving a description of, whatso-
ever upon this Subject, by Dissection,
shall offer it self as an Object of our
Senses.

THE

THE
ANATOMY
OF THE
BRAIN.

CHAP. I.

Of the *Anatomy* of the *Brain*.

THE topmoſt part or *Olla* of the *Cranium* being removed, the firſt part of the *Brain* that comes in view is the *Dura Mater*, which, with the ſubjacent *Pia Mater*, is accounted only an *improper part of the Brain*, ſtrictly ſo called, however of great uſe in many reſpects to it.

'Tis by *Spigelius* and other Anatomiſts reckon'd, and I think not undeſervedly, the thickeſt and hardeſt Membrane of the whole Body, encloſing the whole Brain, properly ſo called, ſomewhat loſely, ſticking al-

B   moſt

moſt inſeparably to the *Baſis* of the
Cranium, and to the top and ſides, under the Coronal, Sagittal, and Lam-
doeid Sutures,very faſt by the *Sinus's*
whoſe deſcription will come in another place.

In ſome places of the upper part
of the *Cranium*, which on each ſide
of the Sagittal Suture or Vertex are
called *Oſſa Bregmatis*, it adheres not to
the Bone,notwithſtanding the poſitive
Opinion of *Van Roonhuyſe*, in his Let- Roonh.
ter to *Du Foy*, to the contrary, who P. 149.
for that very reaſon would fain take
away in a great meaſure the uſe of the
Trepan and *Treſoyne*, and altogether
the uſe of the Inſtrument called *De-
cuſſorium*, which skilful Surgeons
do often make uſe of to make room
for the diſcharge of ſubſided matter
below the fractur'd place in many
Accidents of the Brain.

'Tis very diſcernably double, as
Columbus and ſeveral others formerly, Col. p. 348
and *Vieuſſenius* lately, have obſerved, Vieuſſ. p. 3.
having very ſtrong and large Fibres
on the inſide, but very ſmall, and
hardly viſible, on that ſide next the
Skull ; as appeared to me, after having firſt let it lye a little time in boi-
ling or at leaſt very ſcalding Water.
But

But as to the diftribution of the
double fort of Fibres on each fide this
Membrane, I could not by any means
find them agreeing with the defcription *Vieuffenius* hath given of them, as
running in an oblique femicircular
manner, externally from before backwards, and in the fame figure internally
from behind forwards; but far otherwife, on the infide, where they are
very ftrong, they feem manifeftly to
have three originals from the top part
of the *Proceffus Falcatus*, before, behind, and in its middle; thofe before
running in a curved manner backwards, half the length, and a great
width of the *Dura Mater*, and thofe
behind running after the fame manner forwardly with this difference,
that a great number of them bend
foon after their rife from that procefs
in a kind of a femilunary way to it
again a little on this fide the rife
of the middle *Series* of Fibres, others
of them making a bigger arch after
having ftretched themfelves wider
upon the *Dura Mater*, bend back
again to, and terminate in the Falx a
little beyond the rife of the aforefaid
middle *Series* of Fibres.

Thofe from the middle part of the
Falx run backwardly, but lefs curved
than the reft, terminating as the Fibres which arife backwardly do, at
fome diftance from the Procefs in
the inward Superficies of the *Dura
Mater.*

As to thofe belonging to the exer-
nal fide or fecond *Lamina* of the *Dura
Mater,* they are extream fmall and
obfcure, running from behind forwards.

Befides thefe, there are no lefs remarkable ones belonging to the Falx
it felf, of two forts of Orders, the one
running ftreight about half the length
of it, on its upper part, from before
backwards, the other tranfverfe, from
the inferiour or fifth *Sinus* to the fuperiour or third, on the hinder part
of the Procefs, and are moft confpicuous there, as the other are towards
its foremoft part.

As to the Ufe of thefe Fibres, it
may be remembred that this Membrane confifts of two *Lamina's,* between which the Veins which reduce
the Blood from the Arteries, which
furnifh the whole Brain with it, run
for fome fpace after the manner of
the Ureters in the Bladder, in large
                    Trunks,

Trunks, before they enter the *Sinus* ; so that the Fibrous Conſtitution of this Membrane here, where the Blood-veſſels are largeſt (together with the curved entrance of them into the *Si-nus*, eſpecially in an erect poſition of the Body) do the office of Valves, ſupport the weight, and promote the aſcent of the Blood. But that which is moſt conſiderable, is this, That if the inward *Lamina* of this part, which makes the inferiour and lateral part of the *Sinus*, was not in ſome mea-ſure furniſh'd with additional ſtrength on this ſide ſuitable to that which it hath on the other, by reaſon of its coheſion to the Skull, the Blood which is continually running through it with no ſmall rapidity, eſpecially in great plenitude of the Veſſels or preternatural Ebulitions, would fre-quently burſt out, or at leaſt cauſe ſuch diſtentions as could not but be very injurious to a part ſo very exquiſitely ſenſible ; yet notwith-ſtanding, tho' Nature ſeems plainly to have made a double proviſion againſt ſuch Accidents, by the tranſverſe Li-gaments within the *Sinus*, and theſe ſtrong and numerous Fibres without, I have rarely open'd any ſtrangled

B ₃   Body,

Body, where fome fuch Rupture, or at leaft Diftention, hath not hapned.

This Membrane hath plenty of Nerves from the foremoft Branch of the fifth Pair, and is thereby made very fenfible, fo that from any moleftation given it by the ill *Crafis* or undue motion of the Blood, it becomes accordingly affected. And as the various diftribution of Fibres before defcribed ferve in a natural eftate to give a kind of fpringinefs to the Veffels, whofe Coats are extended by the Blood as they run between the *Laminæ* of this Membrane, to the end the fame may be the more readily circulated through them ; fo in a preternatural eftate, no doubt, they are fubject to Spafms, which may retard the courfe of the Blood in fuch fort, that in fome kind of violent Headachs, where the Membrane is affected through overfulnefs of Blood, and particularly in thofe which are wont to proceed from Vapours (fo called) or Convulfive Motions of Nervous parts, we often obferve a fixed ruddinefs in the Face, attended with a kind of ftiffnefs and forenefs in the Eyes, proceeding doubtlefs from a ftagnation in fome meafure of the Hu-

humours in thofe parts, through the too flow paffage of them into the reductory Veffels or *Sinus's*. And to this preternatural affection of the *Sinus's* may certainly many other ill Symptoms of the Brain be imputed, and not to any irregular *Syftole* and *Diaftole* of the Membrane it felf, occafion'd through any convulfive or paralytical ftate thereof, as that curious Speculatift Dr. *Mayow* hath affirmed, feeing not any living Diffection hath ever been found to give Authority to any fuch Hypothefis.

Mayow, Tr. 4. p. 49

Firft cefs of Dura ter. It hath two Proceffes, the firft of which arifes from that part of the *Os Ethmoeides,* called *Crifta Galli,* and is extended from thence backwards, as far as the concourfe of the four greater *Sinus's,* commonly called *Torcular Herophili,* in the figure of a Sicle, whence it hath that denomination of *Falx,* and by reafon of the ftrict connexion it hath by certain Membranous Fibres with the *Cranium* in thofe places which are immediately under the Sutures, and with the Brain it felf, by the intervention of the *Pia Mater,* (to which it is joyned both by the intervention of large Blood-veffels, propagated thence to the longitudinal

and

and lateral *Sinus's,* and certain carnous Adnaſcencies, as it deſcends down
betwixt the two Hemiſpheres of the
Brain, and afterwards at its approach
to the back of the *Corpus Calloſum,*
(over which that Membrane is looſe-
ly expanded) both by continuity of
its Membranous Subſtance and Ramifications of *Blood-veſſels,* terminating
in the fifth *Sinus,* at the bottom of the
Proceſs, ſo that in a *Diſeaſed Brain*
I once ſaw it drawn up the length of
an Inch from the ſaid *Corpus Calloſum,*
in the exact form of a membranous
thin Production, continued to the
fifth *Sinus* running at the bottom of
this Proceſs,) it keeps the *Brain*
ſuſpended in ſuch a natural conformation, that it needs not, to that internal part by the Ancients call'd *For-*
nix, nor that by *Vieuſſenius* of late
ſubſtituted in the room of it, call'd
Corpus Calloſum, for its ſupport.

    Another Uſe it hath is, partly to
defend the *Cerebellum* from Compreſ-
ſion, to which, by its connexion with
the *Galli Criſta,* it doth not a little
contribute, but chiefly the two Hemiſpheres of the Brain from the like
Injury from each other, upon its various poſition in Sleep or otherwiſe;
                                    and

and therefore is wanting in many other Creatures, as Calves, Sheep, &c. which not only Sleep lefs, but for the moft part in a lefs injurious pofture.

Second *veifs of* *Dura* *er.* The fecond is that which arifing fo forwardly as from the hindermoft Procefs of the Wedglike Bone, which compofes the back and uppermoft part only of the *Sella Equina*; it paffes up betwixt the *Cerebrum* and *Cerebellum*, all the way adhering to the internal Eminencies of the *Offa Petrofa* to the lateral *Sinus's*, by which means not only the *Cerebellum* immediately, as is commonly obferved, but confequently all the Parts from the beginning of the fourth *Sinus*, or the *Glandula Pinealis*, to the laft *Foramen* of the *Skull*, (*viz.*) the *Caudex Medullaris*, with its Appendices the *Nates* and *Teftes*, (which being placed upon the upper part of the *Medulla Oblongata*, make a fort of an *Ifthmus* betwixt the *Cerebrum* and *Cerebellum*) together with the Nerves proceeding out of it, are defended from the injurious preffure of the hinder Limbs of the Brain.

CHAP.

CHAP. II.

Of the Pia Mater.

THE Second Integument of the Brain, commonly called *Pia* or *Tenuis Mater,* by *Galen* and many others, *Choroeides,* from its likenefs in fubftance and ramification of Blood-veffels to that Membrane of the Se-condines call'd *Chorion,* with much more reafon than *Vefalius,* on behalf of the *Plexus Choroeides* it felf ad-vances againft it; was by all the An-cients look'd upon as its only other Integument, being a very thin and pellucid Membrane, co-extended with the Brain it felf, not only in its out-ward but inward ftructure too, as likewife through all its Plicatures, In-terftices, and Cavities, even over the *Corpus Callofum* it felf, tho' loofely, as hath been already obferv'd, not-withftanding the great *Vefalius* af- *Vefal.* firms the contrary: Which Membrane P. 778, alfo a chance cut in pareing the top- par. 2, part of the Brain down to the lateral Ventricles with a Razor, in a *Body* I lately had, gave me an opportunity
of

of fhowing as fair in thofe Ventricles
as the largeſt Membrane of the whole
Body, to feveral who ſtood by, notwithſtanding *Molinetti*, who laughs *Mol.* p.78.
at all that pretend to have found any
fuch thing, affirms the contrary.

But this is to be enquir'd for either
in recent Bodies, or ſuch who have
before death been, thro' fome Difeafes, fill'd with **extravafated** *Serum*, as
Dropſies, *Stoppage of Urine*, fome fort
of *Apoplexies*, or the like : That way
which in want of the other opportu-
nities difcovers it beſt, is the fepara-
ting the *Septum Lucidum* near to its
rife, which is juſt from the *Fornix*,
where it arifes from its two Roots,
near to which place the *Medulla* of
the Brain begins to advance into the
Corpora Striata ; for from thence for
above half way of its paſſage backward toward the hinder limbs of the
Brain, it continues hollow, and, I am
apt to think, is but a Duplicature of
this part, tho' it may be fomewhat
medullary, and therefore, by reafon
of its tranfparency, hath the Name of
Septum Lucidum.

This Opinion of the Ancients, of
its being the only other, and that a
ſingle Integument of the Brain, was
                              equally

equally receiv'd for Truth by the late
two learned and curious Anatomists
Willis and *Vieuffenius*, together with
all the other modern Writers, except
Bidloo and *Bohn*, both which affirm, *Bid.* Tab. 8
they have found another distinct f. 5, 8.
membranous Integument of the Brain *Bohn* p. 333
coming betwixt the other outward
Dura, and inward *Pia Mater,* the one
three hours, the other fifteen days
after death; and by them both recko-
ned the original of the second proper Integument of the Spinal Mar-
row which *Tulpius* first discovered, *Tulp.* cent. 1
and *Vieuffenius* suppofes to be a Dupli- obf. 29.
cature of the *Pia Mater* in that part *Vieuffen.*
only.   p. 143.
                                         par. 2.

Now, that there was a middle
Membrane in fome parts of the Brain,
and particularly at the Bafis of the
Cerebellum, from whence it's continued down to the Spinal Marrow,
conftituting the fecond proper Integument of that part as afore-mentioned, I had long fince obferved; but
whether it be another abfolute diftinct Membrane from that other fubjacent one, by the aforefaid Authors
properly named the *Pia Mater,* and
common to the Spinal Marrow with
the Brain it felf, like as is this other
                                         fe-

fecond middle one too, or only one
and the fame Membrane double, as
confifting of two *Lamina's*, may well
be doubted of.

Wherefore, for fatisfaction concerning this difficulty, I have lately made
the ftricteft enquiry poffible, and that
in a fubject moft likely to afford a
decifion in fuch a Controverfie, and
this was an Human Brain extreamly
hydropical, where there was no Cavity or Interftice, without abundance
of Water extravafated, infomuch that
where ever, according to the natural
conftruction of Parts, there was any
larger than ordinary duplicature of
this Membrane, as there are at the
end of the *Calamus Scriptorius*, betwixt the fuperincumbent *Cerebellum*
and *Medulla Spinalis*, in the *Ifthmus* or
fpace betwixt the *Cerebrum* and *Cerebellum*, upon the Proceffes called *Nates*
and *Teftes*, in the depreffed part alfo
of the Brain, between the beginning
of the Annular Procefs, and the firft
appearance or coming out of the Olfactory Nerves, by *Vefalius* taken
notice of and called a *Procefs of the* ^*Vefal.p.794*
Pia Mater, there was found a great
deal of Water diftending this Duplicature much beyond its natural limits;

mits; fo that by way of confequence,
if thefe Cavities were only Interftices
of two different Membranes diftinctly invefting the Brain, and not a Duplicature only of one and the fame,
the *Water* would then probably have
infinuated it felf betwixt them, and
made them to have appear'd far different from what they did, agreeable to
what it hath often been found to do
in fome *Dropfies of the Belly*, where
the Water hath been found fo to have
divided or parted the double Membrane of that Region call'd *Perito-
neum*, as to have render'd it capable
of containing the quantity of fifteen
Gallons of Water, and upon a difcharge of the fame after death, by
cutting the external *Lamina* of that
Membrane, the other inward one being yet (unknown to the Diffecter)
left whole, to have impofed upon the
Spectators, and thofe very fagacious
ones, fo as that at firft fight, till after
having recollected themfelves, and *Job Metcr.*
divided the other fecond *Lamina* too, Obf. 52.
they thought the Bowels of this part
to have been wanting; but contrary
to this Event, in this Subject I found
this Membrane entire, and free from
any divulfion throughout its whole
cir-

circumference, excepting the places
afore taken notice of. However, fuppofing the like conformation here in
this with the Membranes of the other
parts, I attempted to divide it, and
did fo fuccefsfully in many parts of it,
but moſt readily in the beginning of
the fuperficial Plicatures of the cortical part of the Brain, where there are
naturally fmall Interſtices, betwixt
which many of the Blood-veffels creep
into and immerge themfelves in the
cortical and medullary parts thereof:
So that I think there cannot remain
any further fcruple of its being only
a double, and not two diſtinct Membranes of the Brain.

Bidloo very truly obferves this firſt
or middle Membrane, by him fo called, by me only the firſt, or one *Lamina* of a double Membrane, to be
thinner than the *Dura Mater* above it,
and thicker than the other Membrane
or *Lamina* under it ; which laſt moſt
properly it is that infinuates it felf
through all the clofe Plicatures of the
Brain, and that, as by frequent infpection I have often obferved, not in a
continuous, but rather retiform contexture, and fo, by fuch as love hard
words, or terms of Art, may be
call'd

called after the fame name of that Membrane invefting the cryftalline Humour of the Eye, *Arachnoeides*.

The Advantages accrueing to the whole through fuch a difpofition of this part, as hath already been obferved, are very confiderable, inafmuch as that thereby firft of all it becomes not only an Integument of inclofure, on behalf of the Brain, and the Bloodveffels belonging to it in general, but of expanfion for Strength too, where the peculiar ftructure of Parts, in fuch places as were before mentioned, require it.

As to the firft, the Brain is not only kept more warm, clofe, and compact, and better defended on its depending part from the afperity of the Bone it lies upon, but the Veffels hereby more ftrongly fupported, and it felf fecured from being broken or torn, whilft between its duplicature they climb up into the Brain, whofe delicate tender Fibres muft otherwife of neceffity have fuffer'd violence by the largenefs and pulfation of the Arteries, together with the weight of them, and the other reductory Veffels, ~~from which~~ the *Sinus's* meet them.

Nextly,

Nextly, as it is an Integument of Expanfion in the places before mention'd, that tender fmall part the *Infundibulum*, where it quits the Brain, in order to its paffage into the *Glandula Pituitaria*, by the circumtenfion of this outward *Lamina*, is fortified upon any violent Accident from difruption, and the Brain and *Medulla Oblongata*, in thofe places where they are only loofely contiguous, are better preferved in their natural due connexion ; all which Advantages, inafmuch as they may more reafonably be afcribed to one double Membrane than two fingle ones, tho' of the like ftrength when joyned faft together, may not unreafonably be thought to argue for the duplicature of this Membrane exclufively, to the introduction of a third or new one.

Laftly, as to what concerns the *Glandes* and *Plexus's* which Dr. *Willis* affirms to be fcatter'd all over this *Will.* p. 26. col. 1. Membrane ; as to the former, I could never fee them, but I have feen the external Superficies of the cortical part of the Brain, in ftrangled Bodies, appear glandulous very plainly, through this tranfparent Integument, which upon bare infpection,

C   with-

without further enquiry, might eafily impofe upon the lefs cautious Spe-&ator.

As to the latter, the *Plexus's*, and diftribution of Blood-veffels from them, after a feparation of the ferous grofs part of the Blood in the afore-mentioned fuppofed *Glandules*, (according to that learned perfon's conjecture) into the fubftance of the Brain, in order to produce the finer Animal Spirits; I cannot but look upon it altogether conjectural, till fuch time as not only the *Glandes*, but their excretory Ducts alfo, together with the Emunctories where the fuppofed excrementitious Juice is eliminated, (lymphatick or reductory *Glandes* (if they could be found) never having been by Nature defigned to any fuch ufe) be firft difcovered.

Blood-vef-fels of the Pia Mater. This Membrane hath Blood-veffels of two forts.

Of the firft are thefe properly belonging to the Brain it felf, which, as it hath already been obferv'd, it doth as it were conduct through its Duplicature, in their paffage allowing them thereby the opportunity of growing extreamly fine, after many ferpentine twinings towards their capillary Ex-tremities,

tremities, before they are protended *Bid. Tab.*
into the Brain it felf, and thofe are 8. f. 5. l. M
chiefly fpread all-along upon the un-
der or fecond *Lamina* of this Mem- *Ib. l. G.*
brane.

The fecond are thofe which belong to this part it felf, for its own
nourifhment, and thefe I found upon
diligent infpection, whilft I feparated
its fecond *Lamina* fpread plentifully
upon the infide of the outermoft or
firft *Lamina*, and both thefe you
will find very well delineated in the
places quoted in *Bidloo*.

This Duplicature is alfo very plainly communicated to all the Nerves
both within and without the *Cranium*,
making by its outward *Lamina* a fecond Integument under the firft from
the *Dura Mater* to the whole *Fafciculus* of Nerves, and a third by its inward *Lamina*, which yields an *involu-
crum* or covering to each fingle *Fibrilla*, which collectively make up
the whole Nervous Body it felf, thro'
the admirable finenefs of which Membrane invefting thofe medullary Fibrils, altogether infenfible of themfelves, it happens there is fuch a
nimble confent betwixt part and part,
and betwixt all and the Brain it felf.

C 2   CHAP.

C H A P. III.

Of the Veſſels belonging to the Brain *in general.*

THE Veſſels belonging to this part in common with the reſt of the Body, though in reality but one continued Canal variouſly modified, yet, through the diverſity of Fluids they contain, go commonly under the denomination of *Arteries, Veins, Sinus's,* and *Lymphæducts,* and not without good reaſon, perhaps, the *Nerves* may be in ſome ſence of the ſame kind too.

The two firſt of theſe may, with relation to their different diſtribution, be deſervedly conſider'd in a two-fold reſpect, either as they belong to the firſt Integument of the Brain, or the Brain, properly ſo called, it ſelf.

The Arteries therefore belonging to this part called *Dura Mater,* or *firſt Integument,* are three fair Branches on each ſide.

The

The firſt and foremoſt of which
are ſent out from the Carotid Artery, whilſt it remains in the fourth
10. 2. hh. hole of the *Cranium*, and are propagated chiefly through the fore-
moſt part of the bottom of the *Dura
Mater*, as in the Figure delineated,
but greatly miſtaken by Dr. *Willis*, Willis p. 2.
perhaps taking it upon truſt from col. 2.
Wepfer, equally with himſelf there- wepf. p. 105
in miſtaken ; who deſcribes it for a par. 2.
ſmall branch of the Carotid Arte-
ry, that runs betwixt the two firſt
Lobes of the Brain, which inſtead
of coming out of the Bone of the
Forehead, as he would have it, goes
into it without lending any branches
to this Membrane at all, being truly
delineated and deſcribed by the
aforemention'd accurate *Vieuſſenius*. Vieuſſ. Tab.
And that this Artery was not on- 17. dd, bb.
ly miſtaken by, but unknown to the p. 32.
aforeſaid *Wepfer*, is plain, ſeeing he Wepf. p. 101
ſays, that from the very ſtyliform
Proceſs, where the Carotid Artery
does indeed enter the long Canal, to
the place where it perforates the
Dura Mater to enter the Brain,
there is not one Branch ſent out
from it ; which Error, by injecting
with Wax, which keeps longer in,
C 3　and

and shews the Vessels much better
than small tinged Liquors, had very
easily been avoided.

FIG. 2. ii. The second Branch of Arteries
ascend into the *Dura Mater* by the
sixth hole of the *Cranium*, together
with a Branch of the internal Jugular Vein, and are dispersed laterally
all over the fore-part of this Membrane, as far as the very *Sinus Longitudinalis*, (which nevertheless it
enters not, as there will be occasion to take notice of hereafter)
as in the Figure delineated.

The third Branch of Arteries
climb into the *Dura Mater* by the
eighth hole of the *Calvaria*, together with a small reductory Branch
of the Vertebral Vein, where the
FIG. 2. kk. lateral *Sinus's* enter the internal
Jugular (which occasion'd the Ingenious *Highmore* erroneously to be-
lieve it enter'd the very lateral *Si-* *Highmore,*
nus it self) and the eighth pair of p. 206.
Nerves pass out of the *Cranium*, par. 1.
which passage of this Artery is not
hitherto described by any that I
know of; neither have I ever seen
it figured, but in *Vieussenius*'s first *Vieuss*.tab. 1
Cut, and there but very faintly. kk.

It

It arises from the external Bran- *Vieuss.* tab.
ches of the Vertebral Artery, accor- 8. f. 1. c.
Barthol.
ding to *Vieuffenius*, but *Bartholine* p. 431.
makes it to be a flip of the Carotid par. ult.
Artery, calling it the leffer Branch
thereof; wherein he is miftaken.

The Veins
of the Dura
Mater.

As to the Veins, *Riolane*, and af- *Riol.* p. 252
ter him *Willis*, feems to fay this par. 2.
Will. p. 2.
Membrane hath none; for tho' the col. 2.
latter hath this obfcure expreffion of par. 6.
them, *Tam crebris Venarum propagi-*
nibus quam Arteriarum nufquam con-
fita eft; fpeaking of the *Craffa Me-*
ninx, by which we might guefs he *Will.* p. 22.
thought it had fome, yet in another col. 1.
place he plainly fubftitutes the *Si-* par. 4.
nus's for the reductory Veffels, as
well on behalf of this Membrane as
the Brain it felf; as appears plain
enough in the Page noted.

Vieuffenius indeed allows Veins *Vieuss.* p. 3 t
to this part, and fays, they all-along par. 3.
accompany the Arteries, and after-
ward terminate, according to *Veflin-* *Vefling.*
p. 210.
gus, in the internal Jugular; yet in *Vieuss.* p 4.
another place he fays, fome of the par. 2.
Venal Branches difcharge the Blood
into the *Sinus Longitudinalis*. Which
laft is a flat contradiction to the
place foregoing, inafmuch as in that
he fays, they accompany the Arteries

C 4 all.

all-along after the fame manner of
diftribution or ramification ; which,
if fo, who fees not that they muft
needs grow capillary towards the
Sinus, and confequently be uncapable of reducing the Blood into them,
all reductory Veffels being always
capillary in the place from which,
and not to which, they bring that
which they contain.

Now therefore, neither what the
one nor the other fays can poffibly
be true ; for, as to the former the
learned Dr. *Willis*, if his Affertion
was good, it muft of neceffity follow, that all the Arteries difperfed
thro' this Membrane muft terminate
in fome of the *Sinus's*, otherwife
there will want a reductory Veffel ;
the firft of which is contrary to ocu-
lar demonftration, the laft to common reafon.

As to *Vieuffenius* the latter, befides what hath been already faid
againft him, if what he fays in the
place aforecited be true, that the
Veins of the *Dura Mater* run concomitantly along with the Arteries,
then they muft of neceffity anfwer
the ends of other Veins throughout
the whole Body, in reducing the
Blood

Blood adduced by the Arteries, unlefs the Arteries they accompany difcharge their Blood into the *Sinuffes*,
(which, as hereafter fhall be fhown,
they plainly do not) for otherwife,
feeing they both grow capillary in
their afcent from the Bafis of the
Cranium, they muſt neceffarily be
both adductory Veffels, than which,
by the Laws of Circulation, there
can be no greater an Abfurdity.

Wepfer not knowing of thefe Veins,
was forced to think, and confequently to affirm, That the Arteries leave
the *Dura Mater* in their extremities,
and terminate in the *Pia Mater*, and
fo have their Blood reduced by the
Veins there; but this is evidently not
fo to the Eye of any who heedfully
feparates this Membrane from the
other.

Before therefore I proceed to the
defcription of the Blood-veffels belonging to the Brain it felf, which by
the exactnefs of method I ought to
do, I hope it may be pardonable, if I
make a fhort enquiry after the unac-
cuſtom'd diftribution of Blood-veffels
Nature hath furnifh'd the Brain in
general with, and the Reafons of
its procedure therein.

The

The Truth then concerning this
affair, is, That contrary to what hath
hitherto been obferved, the Blood-
veffels belonging to this part in general, as hath already been obferved, are
of two forts, the one belonging to
the Brain it felf, the other to its outmoft Integuments.

Now, as to the firft, 'tis obferva-
ble, that the Veins enter not the
Brain, nor run concomitantly, like as
in other parts of the Body, with the
Arteries, (the carotid entring at the
fourth hole in the Bafis of the Skull,
and the internal Jugular at the
eighth ; the Vertebral Artery at the
laſt and largeſt hole of the Skull, and
the Vertebral Vein at the ninth
(which *Vieuffenius* miftakenly calls *Vieuffen.*
the tenth) thro' which it runs into p. 163.
the internal Jugular, at that Veins par. 3.
entrance into the round hole at the
bottom of the Skull, under the Styli-
form Procefs, where the *Sinus Lateralis* meets it) where after having advanc'd into certain venous produ&i-
ons called *Sinus's*, they defcend from
thence in large Trunks, growing capillary all-along in their paffage till
they meet the Extremities of the Arteries, and are indeed no other than
                                    meer

meer Branches of the *Sinus's*, and
confequently I look upon the *Sinus's* themfelves no other than large
Veins.

The common reafon all modern
Authors give for this different diftribution of Blood-veffels belonging to
the Brain, from the other parts of
the Body, is, that it may receive an
equal warmth at the top as at the
bottom, as being thereby very much
affifted in the production of Animal
Spirits in an equal proportion all
over ; and that it is fo may very
well be granted : but, that Nature
had yet another provident Intention,
will be as evident, if we confider, that
if the Veins had afcended with the
Arteries thro' the holes in the bottom of the *Cranium*, upon all great
Ebulitions of the Blood, the pulfation
of the Arteries would in that Stricture
of the Veffels made by the Bone, of
neceffity hinder the freedom of its
return by the Veins, and confequently occafion a ftagnation of Blood
through the whole Brain, to the ut-
ter fubverfion of all its faculties, nothing being more certain, than that
upon any confiderable abatement of
circulation there prefently happens
by

by way of reſtagnation, a ſeceſſion of
the watery and thin from the more
groſs and red part of the Blood.

The other way of the **Veins** entring the Brain (*viz.* thoſe appertaining to its outward Integument, one
at the ſixth hole of the Baſis of the
Cranium, the other at the eighth, as
aforeſaid) is, their aſcent with the
Arteries after a quite different manner from the former, even to their
capillary **Extremities**; a manifeſt in-
dication that they ſerve for the reduction of ſo much Blood from the
Dura Mater as the aforeſaid ſort of
Veſſels, the Arteries, have brought
thither ; and although by reaſon of
their ſmallneſs Nature ſeems not to
have been ſo ſollicitous in avoiding
the Inconvenience ſuppoſed to have
follow'd, upon the Artery's entring
the ſame hole with the Veins, taken
notice of in the preceding Caſe,where
they are very large,and conſequently the Effect might prove much
more injurious, yet Nature hath not
been wanting in providing a Reme-
dy againſt it ; as will plainly appear
in the following Pages.

From

From this manner of their entring
the Brain at the fame inlet of the
Skull with the Arteries, may, for
ought I know, be very rationally
accounted for that violent trouble-
fome Noife which many, in Diftem-
pers arifing from the turgefcency
of the Blood, caufing a preternatural
beating of the Arteries, do fo much
complain of; a Symptom happening
from the Stricture before mention'd
which the unyielding circumference
of the Bone occafions upon the different Blood-veffels entring at one
and the fame Foramen, to which
effect alfo the nearnefs of the *Os Petrofum*, through which the Hearing
Nerves do pafs to this hole, which is
in that part of the Wedglike Bone
that joyns to, or is conterminous with
it, does not a little contribute.

To the fame caufe, in fome meafure
doubtlefs, may be afcribed the fre-
quént Headachs happening in Fea-
vers, the Artery then fo fwelling
and compreffing the Vein againft the
edges of the Bone, that the Blood cannot be returned back through it in
a due proportion, and confequently
by its ftagnation the Membrane becomes inflamed and painful.

So

So that conformable to what
hath already been taken notice of
concerning the wife contrivance of
Nature, in ordering the different
diftribution of the Blood-vefiels, fo
as to avoid the Inconveniencies
which might accrew to the Brain
by compreffion of the reductory
Veffels, occafion'd through their
entrance at one and the fame hole
with the Arteries; it feems very
much worth our obferving, that befides the Veins of the *Dura Mater*,
which enter the *Cranium* together
with the Arteries, as hath before
been mention'd, there are alfo feveral others belonging to this Membrane, having their rife at, and their
defcent after a very remarkable
manner, from a Vein hereafter to be
defcrib'd on each fide of the Longitudinal *Sinus*, as you may fee in
FIG. 4. the Figure, and confequently muft
dd,nn,&c. grow capillary in their defcent down
from it, after a quite contrary manner to the other; and thefe do vifi-
bly inofculate with fome of the Extremities of the aforefaid capillary
Arteries, after the fame manner as
thofe larger Veins belonging to the
Pia Mater do with the Arteries
be-

belonging to the Brain and it, by
which means it fo falls out, that a
confiderable part of that Blood
brought up by the *Merinx Arteries*, is carried back by thefe Veins,
to the end that, efpecially in all
preternatural fwelling of the Blood,
the inconvenience of Compreffion
and all its ill confequences happening, by reafon of an overfulnefs of
thefe Veffels, may be in a great
meafure avoided.

CHAP.

C H A P. IV.

Of the ~~Hints~~ *Vessels*, belonging to the
Brain *it self*.

AFTER this fhort digreffion, by
order of Method, the Blood-
veffels belonging properly to the
Brain it felf, fall under confideration.

The curious Anatomift *Malpighius*, Malp. *de*
in his Letter to *Fracaffatus*, fays, they Cereb. p. 6. par. 2.
bear a third proportion to thofe of Dr *Cort.*
the whole Body ; and for what rea- Cereb. p 81
fon, feeing the part it felf bears not par. 2.
the fame proportion to the whole, it
is fo, it will be worth our while to
enquire hereafter.

Thefe are either Arteries or Veins.
The former go under the name of
Carotid and *Vertebral.*

The firft of which, after a curved
paffage (which is very well expreffed
in a Fig. of Dr.*Willis*) from the place *Willis* p. 29.
where it begins to enter the Bafis of Fig. I.
the *Cranium* (which is from the Styli-
form Procefs of the *Os Petrofum*) to
the place where, on the infide, they
pafs through the *Dura Mater*, and
afcend into the Brain, (which is at the
fore-

+ Which laſt have already been treated of

foremoſt internal Proceſs of the *Os
Cuneiforme*) there is very near an inch
and an half diſtance. I ſay, after this
crooked paſſage into the Brain, they
are propagated quite through its ſubſtance, having firſt diveſted themſelves of that thick Coat borrowed of
the *Dura Mater* during their ſtay in
the paſſage aforementioned ; but not
without the mediation or intervention of the *Pia Mater*, which Membrane all the Branches of the aforeſaid,
as well as the Vertebral Artery, more
or leſs firſt prop themſelves upon, before they enter on and diſperſe themſelves through the ſubſtance of the
Brain it ſelf, and is very finely expreſ-
ſed in a Cut of *Placentinus*, at the end *Spig.* p. 179
of *Spigelius*; inſomuch that *Molinetti* *Mol.* p. 77.
(with whom alſo agrees *Marchetti*) p. 191. *Marchetti,*
looks upon it as only a production par. 5.
of thoſe numerous Veſſels ; whereas
all thoſe little ramifications both of
the Carotid and Vertebral Arteries,
viz. thoſe from the carotid Artery,
which as ſoon as it gets through the
Dura Mater, and parts with its bor- a *Vituſſen.*
rowed Coat, are ſent to the a *Infun-* p. 35. par. 1.
dibulum, b *Olfactory*, and c *Optick* p. 34 par. 6.
Nerves, together with thoſe other b Tab. 17.
of the Vertebral Artery which accom- e *Ib.* g g.

D   pany

pany the d third, e fourth, f fifth, g sixth,
h seventh, i eighth, k ninth, and l tenth
pairs of Nerves, inasmuch as they enter not the Brain it self, are altogether
exempt from that Membrane; any of
which now-mention'd Blood-vessels
you either find delineated in *Vieusse-*
nius's 17th Table, or mention'd in
some other place of his Book, by
those Directions here placed in the
margin; all which, tho' existent in
Nature, are neverthelefs there painted
too ftiff and formal (I am afraid by
guefs) inasmuch as that without an
injection of Mercury (except those
two which belong to the *Olfactory*
and *Optick Nerves*) they do rarely
come to fight in any form at all,
Wax being over grofs a body to enter such minute Veffels as those are;
whereas by an injection with Mercury I find fcarce any Nerves but what
hath fome such fmall ramifications of
Blood-veffels in them.

To go about to defcribe diftinctly
the whole ramification of Arteries
through this part, which as was before noted, is here more remarkable
for number and fize than in any other
part of the Body, would not only be
to do what in a great meafure hath
been

d *Vieuffen.*
p.35.par.t.
e Ibid.
f Tab.17. p
p.35.par.t.
g Tab. 17.
T T. Tab.4.
h h. p. 35.
par. 1.
h Tab 4. hh.
i Tab. 17.
Fig. 2.
Tab.4. h h.
k *Ib.* Fig. 2.
l Tab. 4.
h h.

been already done by *Vieuſſenius*, in
his ſixth Chapter, but ſeem to have
alſo in it much more of oſtentation
than uſe.

I ſhall therefore only take notice
of ſuch propagations of them, as are
either remarkable for magnitude,
ſome curioſity of Structure, or uſeful
deſign of Nature.

And of this ſort may well be eſtee-
med the Vertebral Artery, next after
the Carotid, which hath already been
deſcribed, as entering the Brain at
the laſt and largeſt Foramen of the
Skull, contrary to what Dr. *Willis*, *Willis*,p.29.
and before him *Wepfer*, affirms, col. 1.
coming thither on each ſide out of par. 2.
the hole in the tranſverſe Proceſs of par.1. ibid:
the firſt *Vertebra* of the Neck, after *Wepf.*p 112.
a very remarkable curved manner, as *Low.*Tab 4
Fig. 1. EE you ſee in the Figure, (and by no
means like to the delineation and deſcription given by Dr. *Lower* and
Dr. *Willis*,) aſcending laterally
upon the *Medulla Oblongata* as far
as the beginning of the *Proceſſus
Annularis*, where they meet together
in one ſingle Trunk continuing ſo the
length thereof, by *Vieuſſenius* call'd *Vieuſſen.*
Arteria Cervicalis, after which they Tab. 4.
either ſend forth two Branches, or re- bb.
ceive two from the carotid Artery, by

D 2   means

means whereof there is a communication betwixt thefe two large Blood-veffels, and that of great ufe and benefit to the Brain, for by this means it happens, that if even three of the four great *Arteries* which furnifh this part with Blood, were totally obftructed, there would yet be a way left for a competent fupply from the other unobftructed fourth. Thefe I call the *Communicant-branches*, very ill painted in *Bidloo*'s ninth Table, but very well in *Vieuſſenius*'s fourth; as may plainly appear here in the Figure taken exactly from Nature it felf.

FIG. I. dd

Vieuſſen. tab. 4. b b.

The ftructure and fmallnefs of thefe Arteries feem to fuggeft two, yet further, provident Intentions of Nature.

The firft is the fame it hath exprefſed in feveral other places, as in the afcent of the Blood by the Carotid Arteries, both which enter the Brain in a crooked line, the firft at the fourth hole of the Bafis of the Skull, the fecond from the hole in the tranfverfe procefs of the firft *Vertebra* of the Neck, after the manner already in both places defcribed. So in the like manner here, by the narrownefs of thefe Branches, the Blood is in a great meafure retarded in its motion

motion to the carotid Artery, and by confequence to the Brain it felf, which, for Reafons hereafter to be given in defcribing the *Sinus's*, would otherwife be in great danger of being overflowed with extravafated and reftagnant Blood.

The fecond is, a forcing the Blood more plentifully into the Spinal Artery, with which, tho' through the conical ftruĉture of the Arteries in common it cannot be altogether unfurnifh'd, yet by its perfectly-reflexed pofition, would have it very fcantily, were it not that by reafon of the narrownefs of the aforefaid *Communicant-branches* betwixt the two great Arteries, the Blood was driven back in a fort of a retrograde motion.

'Tis true, there is a conformation of Arteries fomething like this, tho' not altogether in the mammary and epigaftrick Branches ; but 'tis worth noting, that in both thefe places the main Artery from which thefe Branches fpring is much more taper or *Ibid. p. c a* conical, and the fucceeding exporting Veffels far lefs both in number and fize than thofe of the carotid Artery here, whofe foremoft and hinder lateral ramifications between the

D 3   Lobes

Lobes of the Brain, bear an over-
proportion to the Trunks from
whence they come, and confequently
muft, according to the aforefaid obfervation of *Malpighius*, in his Letter
to *Fracaffatus*, receive the blood
brought thither far more freely and
plentifully.

Befides, the Cervical Artery here
is fo far from being Conical, that being made up of two vertebral Arteries
joyning together, it is much wider than
either of them fingle, as appears plain-

Fig. 1. g. ly in the Figure, and confequently
would have carried away the Blood
forwardly from the Spinal Artery
more freely, had not Nature order'd
the Structure of Veffels after another
manner here than it does in other
parts of the Body, where there is not
the fame neceffity of contrivance.

One more Branch I take leave to
mention only upon the fcore of its
never hitherto having been taken
notice of by any, and that's a fmall
Artery attended with a Vein paffing
through the lateral part of the *Os Cuneiforme*, (which conftitutes the back
part of the *Orbite of the Eye*, juft
under a very little Procefs of that
Bone, (which either by reafon of its
fize

size hath escaped being seen, or inconsiderable use, was never before, as far as I know, thought worth the mentioning;) and this, upon raising the fore Lobes of the Brain, offers it self to the Eye of any heedful Observer.

CHAP. V.

Of the Sinus's *belonging to the* Brain.

A Third sort of Vessels offer themselves next to our consideration, under the general name of *Sinus's*.

These formerly were reckon'd only four, to which *Vesalius* added a fifth at the bottom of the *Falx*, by him only call'd a *Vein*, which tho' frequently found, yet in some Subjects is wanting. *Bourdon* mentions two more at the bottom of each side the second Process of the *Dura Mater*,

Vesal.p.758 Fig. 3 F.

Bourd. p. 195. par. 2.

D 4   under

under the lateral ones, which I never
faw but once, and I am apt to think
with *Vieuſſenius*, are moſt commonly
wanting.

Vieuſſenius deſcribes four more, *Vieuſſ.*
which I find long before taken notice *p. 6. par. 5.*
of, and exactly deſcrib'd by *Falloppius*, *Fall. tom. 1.*
and after him, tho' but rudely, by *p. 114.*
that laborious Collector *Vidus Vi-* *Vid. Vid.*
dius. *p. 117.*
cap. 10.
p. 310.
cap. 11.

I think I can ſhew one more, but
be their number what it will, I judge
it reaſonable to look upon them no
other than Veins, whether we conſider them in reſpect to either Office
or Structure. All the buſineſs is, to
conſider and ſhew for what end they
appear as ſuch large Channels into
which all the Veins of the Brain,
like ſo many ſmall Rivulets after an
unuſual manner do empty themſelves;
and that I will endeavour to do after
having firſt ſhown their ſeveral re-
ſpective ſituations.

The firſt two are called *Laterales*,
Fig. 4. BB. which run within a ſtrong duplicature
of the hinder Proceſs of the *Dura Mater*, down upon the *Os Occipitale* over
the *Cerebellum*, till in their further deſcent, after a tortuous manner, upon
the lower production of the *Oſſa Pe-
troſa*

trofa they wind under them in order
F16.2.GC. to their paſſage out of the *Cranium* at
the eighth hole, common to the .
Ibid. b b. eighth pair of Nerves going out, the
third Branch of Arteries belonging
to the *Dura Mater,* and the internal
Jugular coming in, which is through
Ibid. L. two round bony Cells in the *Os Pe-
troſa,* juſt under the Styloeid Proceſſes
into the internal Jugular Vein, into
which, together with the Vertebral,
all the reſt of the Veins and *Sinus's*
belonging to the Brain diſcharge the
refluent Blood.

F16. 4. The next is called the third or
AA, &c. longitudinal one, from its riſe at the
bony Proceſs called *Criſta Galli,* and
progreſs the whole length of the
Brain to the hinder and ſomewhat
declining part of the occipital Bone,
where it ſeems to be cleft into the
two lateral ones.

Into this third *Sinus* not only the
internal Veins of the Brain it ſelf are
inſerted, but alſo ſome of thoſe belonging to its outward Integuments,
which *Falloppius* firſt, one of the Lu- *Fallop.*
minaries of Anatomy, obſerved; and tom. 1.
after him *Vieuſſenius,* which are by *Vieuf.* p.10.
Wepfer miſtakenly taken for Arteries, par. 2.
who neverthleſs, for ought I know, par 2..

p.82. par.3

Wepf. p 42.

may

may be in the right, in affigning the
overclofenefs of the Pores of the *Cranium* (by what Accident foever happening) thro' which the refluent
Blood is tranfmitted to the *Sinus*,
for a frequent caufe of inveterate ob-
ftinate Headachs.

The fourth, which from its fituation may not improperly be called
the *Internal Sinus*, comes from the
under part of the falcated Procefs,
at that point where it becomes continuous to the fecond Procefs of the
Dura Mater, and a large double Vein
belonging to the *Plexus Choroeides*,
together with the fifth *Sinus*, (when
there is one) enters it at an Interftice
made between the end of the *Corpus
Callofum*, the *Nates*, *Teftes* and *Cerebellum*, from whence having firft paffed over the *Cerebellum*, it at laft arrives with the other three at that place
of union, which from its Author hath
ever fince retain'd the Name of *Torcular Herophili.*

The four others of *Falloppius* and
Vidus Vidius, or *Vieuffenius*, by this
laft called *Superiores* and *Inferiores*,
the dd firft two of which being longer
and narrower, are call'd *Superiores*,
are on the Bafis of the Brain *, arife,
ac-

Fig. 4. C.

Ibid II.

Ibid K.

Ibid g.

dd.

according to him, from the *Recepta-cula Sellæ Æquinæ*, by the fame Author fo named, (hereafter to be defcri-bed, though more truly, from the ᶜᶜ circular *Sinus*, as I hope in its place to make appear, running down from thence upon the internal Procefs of the *Os Petrofum*, and terminating in the *Sinus Laterales*, where they begin to be declive and ᐱ tortuous in their paffage to the internal Jugular.

Ibid. EE

Ibid. ᐱ.

Ibid. ee.

The other two, called ᶜᶜ *Inferiores*, which are much fhorter and wider than the others, defcend from the fame place as the former, between the *Os Petrofum* and *Occipitale*, down to the aforefaid eighth hole of the *Cranium*, where the Jugulars come up into the Brain, and end there.

Another I difcover'd by having firft injected the Veins with Wax running round the *Pituitary Gland* on its upper fide forwardly within a duplicature of the *Dura Mater*, backwardly between the *Dura Mater* and *Pia Mater*, there fomewhat loofely ftretched over the fubjacent Gland it felf, and laterally in a fort of a Canal made up of the *Dura Mater* above, and the carotid Artery on each outfide of the Gland, which by being faften'd

faften'd to the *Dura Mater* above,
and below at the Bafis of the Skull
too, leaves only a little Interftice betwixt it felf and the Gland, thereby
conftituting a Cavity communicating
with the two foremention'd forward
and backward ones, from whence the
abovemention'd four fmall *Sinus*'s do
defcend, by a vifible continuity, on
each fide from a little beneath the
hinder Procefs of the *Sella Turcica :*

Fig. 2. EE and this from its Figure may not unfitly be called the *Circular Sinus.*

Vieuffenius, it may be, faw fome
part of this *Sinus* where the other
four fmall ones enter it, which is at
the hindermoft part of his *Receptacula
Sellæ Equinæ lateribus adjacentia,* fo
called, and from thence thought thofe
Receptacles to communicate with
and to be capable of performing the
office he affigns them, (*viz.*) of
bringing back Blood from the nourifhment of the fubjacent Bone call'd
Cuneiforme, together with the Water
feparated from the Pituitary Gland,
into thefe four inferiour *Sinus's.*

Now, as concerning thefe Receptacles of his, 'tis certain that they are
not any where exiftent in Human
Brains, (according to the defcription
he

he gives of them in the place here noted) feeing both the third, fourth, Pag. 16. two foremoſt Branches of the fifth, as well as its third hindermoſt one, together with the ſixth pair of Nerves, do not only run out of the Brain encloſed in ſo many diſtinct little *Capſula's* or Coverings made of the *Dura Mater*, during their paſſage through that part of the Baſis of the *Cranium* by him call'd *Receptacula*, &c. but even the whole *Dura Mater*, together with its Membranous Productions conſtituting the aforeſaid Coverings of thoſe Nerves, in that place ſticks cloſe to the Baſis of the ſubjacent Bone, (*viz.*) the External Proceſs of the *Os Cuneiforme*, on its under ſide, and to the Carotid Artery (which alſo both above and below (as was before noted) by its borrow'd coat ſticks cloſe to the *Dura Mater*,) on that ſide towards the aforeſaid Gland, leaving no room at all for either Blood or *Serum* to be contain'd there, as he would have it ; tho' in the ſame place which he deſcribes for his Receptacles I have in ſeveral injected Bodies obſerv'd two very fair and large Veins, one coming into the *Cranium* at the ſecond *Foramen* from the

the Orbit of the Eye, (and poſſibly may be a Reductory Veſſel to that part) and ſo climbs up on the ſide of the lateral Proceſs of the Wedglike Bone, almoſt up to the *Circular Sinus*; the other at the fifth *Foramen*, which climbs up upon the ſame Bone till it meet and joyns with the other, from whence they make one ſhort Branch, which enters the *Circular Sinus* very near the place where the two other inferiour ones on each ſide deſcend down from it; which if they ſhould chance to be cut by accident in any enquiry made into that part, might cauſe an appearance of Blood, and thereby become an occaſion of the aforeſaid erroneous Hypotheſis.

Neither is it poſſible (granting there were any ſuch Receptacles as he mentions) they ſhould ſerve to the end he aſſigns, ſeeing the *Glandu-* *Vieuſſ.p.55* *la Pituitaria* is on all ſides encloſed by both the *Dura* and *Pia Mater*; which firſt (notwithſtanding what he ſays to the contrary) is on all ſides of this Gland of a very ſtrong and equal thickneſs ; yea, in that very part where (as hath been before ta-ken notice of) there is a kind of a Chaſe made by a certain duplicature

of

of the *Dura Mater*, conftituting the
foremoft part of the *Circular Sinus.*

And if this alfo was granted, yet
would the manner he defcribes of
the *Serum* or Water getting into thefe
Receptacles (which is by tranfcola-
tion) render his Suppofition very unprobable, feeing 'tis by no means
conformable to the Cuftom of Nature in all other parts of the Body
that Arteries fhould depofe a *Serum,*
or any thing elfe but Blood, (except
what goes for Nourifhment to the
Part it felf) in any Part, without being furnifh'd either with its Excretory or Secretory Ductus, neither of
which was ever pretended to have
been found here.

And as a thorow confirmation of
all this, faid in oppofition to the aforefaid Hypothefis, I fhall only add this,
and conclude, that in feveral Injections made ufe of in order to find out
the ufe of Parts, I never found one
drop of the tinged Liquors on that
fide of the Carotid Artery, where he
hath made the fituation of thefe Receptacles.

The ufe of this *Circular Sinus* is in
common with the reft to reduce
Blood returning from all the adjacent
parts, as the Pituitary Gland, the
Wedg-

Wedglike Bone also, and it may be from the *Rete Mirabile*, which in Brutes is very large, and therefore seems to require the Service of this *Sinus*, either mediately or immediately, for reducing a share of its Blood, seeing the *Glandula Pituitaria* appears no where furnish'd with Veins terminating any where else sufficient to carry off the refluent Blood from this *Plexus*, notwithstanding *Vieuffenius* saith on the contrary it hath no Veins, and therefore is forc'd to have recourse to those small Branches of Veins which accompany the Branches sent out by the carotid Artery, before it perforate the *Dura Mater*, with the Optick Nerves, or those which go to the *Gangliforme Plexus* of the fifth Nerve, or those coming out of the Wedglike Bone, for reductory Vessels to this Part ; but with what probability I know not.

CHAP

C H A P. VI.

Of the Motion of the Brain *and* Sinus's.

TO thefe *Sinus's*, efpecially the *Longitudinalis*, and by way of confequence to the *Lateralis* alfo, moft if not all the Ancients, as well as Moderns too, particularly *Willis* and *Vieuffenius*, have unanimoufly afcrib'd Pulfation, after the manner of Arteries, by reafon of fome Arteries (as they thought) from the *Dura Mater* terminating in them : of the truth whereof being fomewhat doubtful, I refolv'd to make ufe of fuch an Experiment as might remove all future Scruples, and moft fatisfactorily put an end to the Controverfie ; which was as follows.

Vitaffen. p. 14. par. 3.

   I took off the upper part of the Skull of a Dog alive, by which means the *Dura Mater* with its third *Longitudinal Sinus* lay bare to the Eye and Touch, to neither of which Senfes, at firft, either any beating of

E   the

the Membrane in general, or of the *Sinus*, was the least discernable. After some pause, by chance the *Sinus* it self, which I defign'd to have open'd with a Lancet, being touch'd with a cauterizing Iron (which in making the Experiment there was occafion to make ufe of) pour'd out the Blood very violently, and at firft without any very remarkable pulfation, but after fome time difcernable enough, both as to the Blood and Membrane too.

I cut this *Sinus* through almoft the length of it, to fee whether any Arteries (whereof many, according to *Vieuffenius*, which was alfo long afore affirm'd, and that upon Experience too, by the learned *Wepfer*, did terminate in it, and fo occafion its beating.) would difcover themfelves by throwing out their falient Blood, but no fuch Sign appear'd. *Wepf* p 116 par. 1.

After all which 'tis manifeft the *Sinus*'s themfelves have no pulfation, other than what is communicated to them from the fubjacent Brain, which contrary to what *Bourdon* affirms, hath an evident pulfation through the *Board.* multitude of Arteries difperfed thro' p. 196. par. 2.

it

it fo forcible as to create a fenfible *Syftole* and *Diaftole* in its outward coverings.

'Tis worth noting, that while the Blood-veffels are all full, fo as to keep the *Dura Mater* upon its full ftretch, the pulfation is not vifible at all, or at leaft very faintly ; but after a depletion of the Veffels, fo as that grows fomewhat more lax, the beating becomes very vifible, equally in the *Sinus* and Membrane too.

After having made this Experiment I found one Author of the fame opinion, and that is *Falloppius*, who in vindication of *Galen* againft *Vefalius*, his Contemporary, fays, all I have faid upon the foregoing Experiment, and all the great *Vefalius* was able to anfwer in his own vindication in his ingenious Book call'd *Anatomicam Gabr. Falloppi Obfervat. Examen*, falls very fhort of its aim.

As to the Tranfverfe Ligaments
† *Fig.* 4. r. which are in fome places * round, cordal, and in others † broad or
† *Ibid.* x. membranous, in the Longitudinal *Sinus* chiefly, both ferving for Strength and (in concurrence with the cruciform ligamentous Fibres, taken no-

tice

tice of by *Vieuffenius*, on the under
and outfide of this *Sinus*, from whence
the Fibres belonging to the falcated
Procefs aforemention'd feem to have
their original,) Elafticity to this part
for its more vigorous reduction of
the Blood paffing through it, together
with its blind Cavities or Diverticulums ferving to moderate the over-
fwift or violent motion of the Blood;
feeing I find them fo exactly defcrib'd
by *Vieuffenius*, to whom the Reader
may have recourfe, I think their defcription need take up no room
here.

But as to the manner of the Veins
entring this *Sinus*, I find it far different from that which is defcrib'd by
Lower firft, and afterwards by *Vieuf-*   *Low.* fig.4.
                                             h h.
fenius, both whom make them enter *Vieuff.*tab.2
with their Orifices from behind for- D D, &c.
wards, (two or three only excepted
by *Vieuffenius*) and that for fome
other ufeful purpofes than what have
hitherto been taken notice of.

And this is as follows, (*viz.*) About *Fig.* 4.
one half of them (tho' intermixedly) dd, &c.
(but all, after having firft upon their ari-
val at the *Sinus* infinuated themfelves
for fome fpace after the manner of
                                 the

the Pancreatick Duct or Ureters firſt
dd, &c. taken notice of by *Lower*, betwixt the
Duplicature of the *Dura Mater*) from
behind forwards, the other half from
before backwards, as in the Figure.

Now, by this contrivance 'tis plain,
that firſt of all there are made two
contrary Torrents in one and the
ſame Channel, by which means the
refluent Blood, made poor by the vaſt
quantity of its richeſt parts drawn off
as it were into Animal Spirits, thro'
a colliſion of Parts, which by this
contrivance muſt needs fall out, is
preſerv'd in its due mixture, which
when at any time loſt through the
languiſhing of its inteſtine motion or
elaſticity, retards even its circular or
progreſſive motion, which when it
happens but in ſome degree, is the
cauſe of many Diſtempers; and when
altogether, of Death it ſelf.

In the next place the circulation is
at all times not only ſomewhat retarded, and the Blood hinder'd, (together with the help of the bony Cell
at which the internal Jugular Veins
enter the *Sinus's*) eſpecially in an
erect poſture, from deſcending with
that rapidneſs and weight it would

E 3   other-

otherwife have done upon the defcending *Cava* to the Heart ; but alfo
much more fo retarded in a fupine
pofition of the Head, a pofture moft
natural and ordinary for Mankind to
take their reft in, through which contrivance, in concurrence with that of
the Lateral *Sinus's*, (whofe ftructure
is fuch, that in the aforefaid pofture
the Blood is forced to climb upwards
before it can arrive at the place of
its defcent into the Jugular Vein)
there is made a more plentiful gene-
ration of Animal Spirits, one chief
Caufe of the great refrefhment and
vigorous difpofition of the whole
Body we find after Sleeping.

As to the other manner of the Veins
entring this *Sinus*, (*viz.* from before
backwards) it from thence happens,
that in a prone Pofition of the Brain,
a pofture not uncommon amongft
Men, the Blood is help'd forward in
its circulation through the *Sinus* ;
the truth and defign whereof are at
once both evident and pointed at by
Nature from the Structure of this
part (and which therefore fhews the
great ufefulnefs of Comparative Anatomy) in Brutes, who by reafon of
                                    fuch

such a Position, which the necessity
of Feeding almost always keeps them
in, have always such a disposition
of this Part, to assist the Blood in
its heavy circulation.

The design of Nature in making
these Channels so wide on a sudden,
in respect to the Branches of Veins
lately treated of terminating in them,
seems to correspond with the conformation of the Parts just now treated of, and with that it had in making the Ramifications of Arteries
afore taken notice of so large and
unproportionable to the Trunks from
which they spring, which is a slower
than ordinary circulation of Blood
through the Brain, in order to make
a still more copious production of the
Animal Spirits so called. Which pro-
fitable Design and End of Nature had
nevertheless been attended with a
very great Inconvenience, (*viz.*) an
extravasation of too much *Serum*, the
usual effect or consequence of a slac-
ken'd Circulation, had it not been
for another provident Contrivance
of Nature in the two Communicant-
branches, betwixt the Carotid and
Vertebral Arteries aforemention'd,

E 4

p. 36. by the narrowneſs of whoſe Channel the influent Blood is in ſome meaſure repreſt in its motion, and an overcharging the Veſſels with Blood prevented.

Theſe *Sinus's* differ in ſtructure one from another, the Longitudinal and Lateral ones having many tranſverſe Ligaments which the other have not, and the Longitudinal having many ſmall Cavities or blind Diverticulums, as aforeſaid, which the Lateral have not; the uſe of them all being for ſtrengthening and defending them from giving way to the violent irruption of Blood into them, againſt which ſometimes notwithſtanding they are not able to defend themſelves ; as I have ſeen in many Skulls ni which the Blood hath burſt open the ſides of the *Sinus's*, and found its way between the Duplicature of it, ſo as even to have made a *Fovea* or Cavity in the *Cranium* it ſelf, as was before noted, one of which I have now by me.

CHAP.

CHAP. VII.

Of the Plexus Choroeides.

THIS *Plexus* is an aggregate Body made up of *Arteries, Veins, Membrane,* and *Glands,* double on each fide, (which hath not before been taken notice of) and confequently having two Originals.

The firſt Original is from the foremoſt Branch of the Communicant Artery, which running backward up betwixt the hinder Lobes of the Brain, (in which for fome part of the way it is immerged, and to which it gives many large Branches) and the *Medulla Oblongata* at length arrives at the Lateral Ventricles, and makes one part of the *Plexus* on each fide.

The fecond Original is from the hindermoſt Branch of that Communicant Artery, which running more backwardly, afcends betwixt the hinder Limbs of the Brain and the *Cerebellum,* till it comes to the *Iſthmus,*
where

where communicating with the firſt
Branch abovemention'd, they make a
reticular broad Expanſion, which co-
vers both *Nates, Teſtes,* and *Glandula*
FIG 5. GG *Pinealis,* and conſtitutes the ſecond
or other part of the *Plexus Choroeides.*

The firſt Branch begins to divide
it ſelf into divers Network Fould-
ings, interſperſed with Glands ſome-
Ibid. 5. what before it enters the Ventricles,
and continues ſuch to its Extremity
on each ſide, where they both under
the *Fornix* wind croſs the third Ventricle into a mutual inoſculation.

* The ſecond begins to aſſume the
ſame ſhape or contexture as ſoon as
it begins to enter the *Iſthmus,* continuing ſuch throughout its entire
abovemention'd Expanſion.

Theſe two on each ſide are joined
together by a twofold connexion,
the firſt is by an Artery running under the *Bombyces,* intervening betwixt
them, which could not be here inſerted ſo as to come in view.

The ſecond is by a production of
the *Pia Mater,* which is extended all
over theſe parts of the Lateral Ventricles, and the third Ventricle which
lyes

lyes betwixt the firſt two parts of the
Plexus forwardly, and down to the
other two hinder parts of the *Plexus*
backwardly under the *Fornix* and
Septum Lucidum ; ſo that whatſoever
Water is tranſmitted out of theſe Ventricles, muſt ſlip down not only under the *Fornix*, but that Membranous Production it ſelf ; from which
kind of ſtructure and poſition of this
Membrane may probably be under-
ſtood how there might happen ſuch
an *Hydrocephalus* as the learned *Tul-* *Tulp.* lib. 1.
pius mentions, in which there was cap. 24
found above two pounds of Water
in one Ventricle, without any at all
in the other : and ſuch another as
Wepfer mentions, where the Water *Wepf.* p. 69
cauſing the *Hydrocephalus* in an Hei-
fer, was found contain'd in a *Cyſtis*,
and that only in the left Ventricle
too : for, ſuppoſing this membranous
production of the *Pia Mater* to be
double here, as it certainly is in all
other places, 'tis not difficult to conceive, that the Water which is ex-
travaſated muſt needs inſinuate it ſelf
betwixt the two *Lamina's*, till by a
continual encreaſe it extends them
into the ſhape of a large Bladder,
                                    ſuch

such a one as the latter found there and drew out with his Fingers ; and that which seems to put out of all Controversie that it was so, is, that in those places, both above towards the *Corpus Callosum*, and below on the Basis of the Ventricle, he found some sort of Asperities as though the Bladder fill'd with Water had been covered with some small Protuberances not much unlike to White Poppy-feed, in those places where it was contiguous to them ; which Protuberances doubtlefs were the small Glands interfperfed quite through this *Plexus*.

How this Diftemper came to be on one fide only, though fometimes it is on both, as you may fee in another place of the aforefaid *Tulpius*, may likely enough be from an Ad-nafcency of both the *Lamina's* of this Membranous Production, in that place where the *Septum Lucidum* finks down from the *Fornix*, occafion'd by fome fmall fort of preffure of the fuperincumbent Brain. Befides thefe Veins, which are very truly deferib'd by *Willis*, I have always found two more meeting the fore-

Willis p. 1a col. 2, par. 1.

foremoft Extremities of this *Plexus*, from between the two firft Lobes of the Brain, where it feems to end under the foremoft part of the *Corpora Striata*, by which it is there fixed and as it were kept in its due fituation : and from thefe Branches are on each fide fent forth many more little ones to the *Corpora Striata*, and feveral other parts adjacent.

To this *Plexus* belong alfo Veins, which from the Extremities of that part of it in the Lateral Ventricles *Fig.5. hh.* begin to come into two diftinct pretty large Trunks, running down thro' the middle of the third Ventricle, as far as the fourth *Sinus*, and there receiving fome Branches from the other hinder part of the *Plexus* fpread over the *Ifthmus*, difcharge the refluent Blood into that *Sinus*.

bid. qq But befides this fort of Reductory Veffels, it hath alfo another, (*viz.*) *Lymphæducts*, which I firft difcover'd in the Brain of a ftrangled Body, and fhew'd to feveral then prefent, running in different ramifications amongft the reticulated Veffels and Glands of this part : Which Obfervation being added to that of the great Anatomift

An-

Anthony Nuck, who in that curious Piece call'd *Adenographia* fays, he faw one coming from the *Glandula Pinealis*, and that his Friend another Anatomift, whofe Name he mentions not, (but I know it was one *Bodivol*, whom I had the Happinefs to be very well acquainted withal, now dead) fent him word, he faw another not far from the aforefaid place ; may be of fufficient authority to evince the real Exiftence of thefe Veffels hitherto fo much enquir'd after, in the Brain as well as in other parts of the Body.

Nuc p.150.

The Glands belonging to this *Plexus* are very many, but very fmall, and their Ufe, according to all the Moderns, efpecially *Willis*, *Duncan*, and *Vieuffenius*, to carry off the redundant watery part of the Blood, but that without ever fhewing by what rational contrivance of Structure it can be done, feeing none of them afcribe a Secretory Duct, which muft always be in readinefs when any unprofitable part is to be difcharg'd.

Since therefore this part is found furnifh'd with *Lymphæducts*, 'twill be no hard matter to conceive the genuine ufe of the Glands, which is, to fepa-

feparate a rich nutritious Juice from the influent Blood, and by the *Lymphæducts* to refund it to the refluent, after the lofs of its nobleft parts left behind in the Brain, in its paffage to the Heart again.

It may alfo, for ought I know, according to the Opinion of *Willa*, ferve to warm its neighbouring parts the Internal Superficies of the Brain, which being purely medullary, hath not fo plentiful a fhare of Blood-vef: fels difperfed through it as the reft, and confequently, to maintain an equality of warmth conducing fo much to the conferving the Spirits in their due vigour and exercife, muft borrow an additional fupply from hence. It is fituated upon the middle of the *Thalami Nervorum Opticorum*, all-along them length way, and, contrary to what *Willis* fays, is, by vertue of feveral Blood-veffels, join'd to that medullary part of the Brain fo call'd, immediately lying under it.

CHAP.

C H A P. VIII.
Of the Rete Mirabile.

NOtwithstanding the Opinions of the late *Wepfer*, *Willis*, and *Vieuffenius* too, (which two last indeed, tho' but now and then, are willing to allow it an existence ~~only~~ in Men, (who neverthelefs, if the Suppofition of *Willis* be true, *viz.* That fuch cannot but be Fools) had better be without it,) together with almoft all the Ancients, as *Vefalius,Columbus,&c.* to the contrary, I have never found this *Rete* wanting, or with any difficulty difcoverable in Men, fpringing from and lying on the infide of each Carotid Artery, in that place of the Circular *Sinus* chiefly which looks into the four abovemention'd inferiour and fuperiour *Sinus's* in the Bafis of the Brain,and in fome meafure alfo the whole length of the *Sella Turcica*, on each fide, between the Gland and the Carotid Artery.

And that it is fo fmall in them with refpect to what it is in Brutes of feveral kinds, is no way furprizing, when confideration is had to the Ufe and Service of it in thofe Creatures, who,

Willis p. 27. col. 2.

who, by reafon of their prone Pofition, would otherwife be in danger
of having their Brains deluged as it
were with an over-great quantity of
the Influent Blood, and of a Rupture
of the Veffels, by its violent ingrefs,
and this Danger fo much the more
threatned by how much the fame
Caufe which brings it into the Brain
with that force is equally as great and
effectual to hinder its proportionable
return ; for the relief of which Inconveniency Nature hath contriv'd a
means of its more eafie and fafe 'defcent into the Brain, by turning that
one large Stream of Blood, (which
through its being penn'd in one
Channel, becomes fo rapid) into many more, (by which means the Carotid Trunk above the *Dura Mater*
in thofe Creatures is very fmall to
what it is beneath, whereas that Artery in Men, *&c.* hath the fame bignefs on both fides that Membrane,)
and they not only reticulated and
contorted for the more flow and laborious (which Contrivance the
Ancients thought was only for a
more exact preparation of the Blood
for Animal Spirits) defcent of the
Blood, but alfo many of them by
their infertion into the *Glandula Pi-*

F   *tuitaria*

tuitaria, attended with small Veins
issuing thence, to take off some part
of the burden too.

This last contrivance of Nature
methinks may be sufficient to render
that Controversie of *Vieussenius* with
Willis (which, before them, was betwixt *Waleus* and *Rolfincius*) the two
latter on each side denying this *Rete*
to have any Veins, very needless ;
seeing that if the Pituitary Gland have
any, which I am confident it hath,
(notwithstanding the positive Assertion of *Diemerbroeke,* in order to serve
his own most unprobable Hypothesis,
to the contrary) as having seen them
plain injected with Wax ; then this
part of the Blood in some of the Branches of the said *Rete,* which are
plainly inserted into the Gland, is
equally capable of being reduced by
those Veins without any necessity of having recourse to those remote Branches *Vieussenius* hath been
forced to seek for, as if it had had
them of its own.

And that to the aforesaid Position
of different Creatures ought chiefly
to be ascrib'd the variety of Magnitude of this *Rete* in several of them,
its size in *Dogs* seems highly to evince ;

in

*Vieuss.*p.46
par. 2.

Diemerbr.
p. 364.
par. 3.

*Vieuss.*p.46
par. 2.

in which, by reafon of their Hori-
zontal Pofition, being neither fo
prone as feveral Brutes who feed on
Grafs, nor fo erect as Man, that
Rete is found fmaller than in the firft,
and larger than in the laft.

Another Ufe it hath been thought
to have, is, to carry off a confiderable quantity of a dull watery part of
the Blood, in order to the production of the finer Animal Spirits; and
this it is thought to effect by means
and help of the Pituitary Gland, betwixt which and it felf there is conftantly obferv'd a great affinity, the
one being either greater or leffer in
proportion as the other is fo, and betwixt which there are in all Creatures, but more remarkably in thofe
where they are both large, a diftribution of feveral Branches coming
from the aforefaid *Rete*. And this is
look'd upon by *Vieuffenius* fo confiderable an office of the *Glandula Pituitaria*, that in thofe Creatures
where it is but fmall, as in Men,
Horfes, Dogs, *&c.* he hath fub- *Vieuf.*p.102
ftituted many, but particularly par. 3.
two Cavities, for that ufe in the
Wedglike Bone, juft under the *Sella
Turcica*, in which he fuppofes that

part of the aforefaid *Serum*, which
by the fmallnefs of the *Rete* cannot be return'd that way, is re-
mitted by feveral little Arteries flipt
off from the Carotid, whilft under
the *Sella Turcica*, terminating in the
two abovenamed Cavities, there either
depofing a part of the *Serum* to be
carried off by a ftrange way he
there mentions, (*viz.*) by two holes,
into the Noftrils, and thence into the
Fauces ; or elfe by certain Veins
meeting them in that place, as their *Vieuff.* p 9.
proper Reductory Veffels, to the par. 2.
Heart.

Now, as to this office of the *Glandula Pituitaria*, I cannot eafily be
perfwaded it is either defign'd for, or
capable of it, till fuch time the Abet-
tors of this Opinion can be able to
fhow me it furnifh'd with an Excretory Duct for this purpofe.

And if they offer, that the Veins
are fuch, I reply, That (befides its
being very unprobable that fo vaft a
quantity of Blood as continually is
brought by the Carotid Arteries to
the Brain, fhould be able to get rid
of any confiderable quantity of its
Serofity, by fo fmall a part as the
Glandula Pituitaria is ;) 'tis not the
ufual

ufual way of Nature to part with
any Share of its Juices out of its
Veffels, when fo unactive and unprofitable as this is, and immediatoly to
receive it in again, feeing it is provided of Emunctories enough to convey it away by.

Moreover, granting (which by no
reafonable means is to be granted)
it were fo as they would have it, yet
neverthelefs, in conformity to Nature's proceedings in all fuch-like
cafes, there ought to be an intermediate paffage by way of a Secretory
Duct, which none hath been able
hitherto to difcover.

And fo far as *Vieuffenius* feems to *Vieuf.*p. 102 par. 3.
be of this opinion, which in one place
he plainly is, making it of fo grofs
and vifcid a nature, as is only fit to
be difcharg'd at the Emunctory of
the Nofe; the fame Reply is fatisfa-
ctory : But when by way of flat contradiction to himfelf he comes to
make the fame grofs Humour a per-
fect fine *Lympha,* the Anfwer is then, *Vieuff.*p. 54 par. 1.
That there is no need of parting
with it beforehand, feeing we find
that Liquor only feparated by the
Lymphæducts of the Brain afterwards.

F 3 Seeing

Seeing therefore there is such an affinity as before mention'd, between the *Rete Mirabile* and *Glandula Pituitaria*, and taking it for granted, that the office of the *Glandula Pituitaria* is not what it hath generally hitherto been believ'd, to the end we may attain a more exact knowledge of what it really is ; it seemeth not altogether immethodical to take that part into confideration in the next place, together with the *Infundibulum*, which laft hath not only as near a relation to the Gland as the Gland hath to the *Rete*, but fuch a clofe communication with it, that it feems in a manner almost impoffible to treat of one independently on the other.

CHAP.

CHAP. IX.

Of the Glandula Pituitaria, and Infundibulum.

THIS Gland is seated in and fills up in a manner all that space contain'd within the *Sella Turcica* (Veſſels only excepted).

'Tis cover'd on all ſides with the *Pia* and *Dura Mater*, excepting that part on its upper Superficies, in which there is a little round hole, by which the *Infundibulum* deſcends ſlopingly into it, being at its entrance inviron'd with a Production of the *Pia Mater*, for its more firm connexion with that part, as was before noted.

But as to the *Dura Mater*, it encompaſſes it after a far different manner than what *Vieuſſenius* hath deſcrib'd, not ſuſpending it in Man as it doth in Brutes, ſo as to hinder it from touching the bottom of the *Sella*, and that foraſmuch as there is not the ſame reaſon for its ſo doing in one as there is in the other, for in

*Vieuſſ.*p. 51 Par. 5.

F 4   -   Brutes

Brutes the *Rete Mirabile* is not only
fituate on each fide this Gland, but
runs quite under its hinder part, by
which one fide of the *Rete* commu-
nicates with the other, a Difpofition
of this Part which *Vieuffenius* was altogether unacquainted with ; whereas
in Man, inafmuch as there is not that
fort of Structure in the one (*i. e.*
the *Rete*) 'tis not neceffary it fhould
be requir'd in the other.

However, in neither one nor the
other is the Reafon which *Vieuffenius*
gives for Nature's contrivance of this
affair of any weight, feeing neither
the *Rete Mirabile*, much lefs the few *Vieuff.* p. 50
fmall Veins belonging to the Bone par 1.
beneath, could poffibly any way be
compreffed by this Gland, though
fuperincumbent, becaufe it is fo
firmly knit to the *Dura Mater*, lying above and upon it, which is fupported by the two foremoft and hindermoft Proceffes of the *Sella Turcica,*
in fuch a manner as is fufficient to
fuftain and keep from preffing upon
any fubjacent part ten times a greater weight than the *Glandula Pituitaria* is.

Moreover, the *Dura Mater* is fo
far from fufpending it from that
Bone,

Bone, that it is, together with
the Gland, fixed to that very Bone
it felf.

The fubftance of this *Gland* is far
differing from that of all the reft,
which I have often upon this account
particularly examin'd ; in confiftence
indeed 'tis the fame with moft of the
Conglobate kind, if not fomewhat
harder, but then being preffed or
fqueezed, it emits much more Water
than any of them.

As to the Conglomerate fort, it
hath not the leaft refemblance to any
of them, and confequently cannot be
fuppos'd, as it hath hitherto been by
all, to carry off any excrementitious
or unprofitable part of the Blood.

Now, if we confider this part, together with the appended *Infundibu-*
lum, we fhall certainly find a conformation far different from any other
part in the whole Body of Man, inafmuch as that which this Gland
receives by the *Infundibulum*, or which
is the fame, what this *Infundibulum*
conveys to it, is not feparated from
the mafs of Fluids by any vifible
Secretory Duct, which in its ordinary method Nature is obferv'd conftantly to make ufe of, whenfoever
it

it parts with any part of the
Blood, whether excrementitious or
reductitious, throughout the whole
compages of the Body.

Nor hath the manner of Nature in
tranfmitting a certain Liquor to the
Gland been lefs abftrufe in carrying
it off from that part again, the reductory Veffels from the Gland being equally conceal'd, as the adductory to the *Infundibulum* ; that way
of Tranfudation , according to the
invention of *Vieuffenius,* being to the
greateft degree improbable, as having
no refemblance to the courfe of Nature throughout the whole Body.

Nay, even a poffibility it felf feems
hardly allowable, if we take but notice of that part in Brutes in whom
its Integuments are extraordinary
denfe, the *Dura Mater,* as he truly
obferves, invefting it clofe on every
fide, (and which he perceiving, and
confequently forefeeing what might
from thence unanfwerably be objected
againft him) was forced to make
them much more than in Men; in
which laft indeed there is feemingly
fome reafon for its being fo, inafmuch as the *Rete* lies in a Duplicature as it were of the *Dura Mater,* on

*Vieuff.*p.52 par. 2.

each

each fide of the hindermoft part of
the *Sella Turcica*, as tho' one *Lamina*
of it was fpread upon the fubjacent
Bone, and the other over the *Pituitary Gland*, (a difpofition contrary
to that in Brutes, as hath already
been taken notice of) but neverthelefs there is no neceffity that it fhould
be fo divided in this place, nor doth
the faid Author ever offer a Reafon
for its being fo, (which looks as
though his Affertion was only a
Guefs) feeing this Membrane can
fend out new Productions as well
double as fingle, as we find in its
two eminent Proceffes before defcrib'd, and *Sinus's* ; agreeable to
what it alfo therefore may and does
do here, where the Integuments of
this part appear plainly to be of too
thick a confiftence to admit of his
imaginary way of tranfudation,which
is manifeft not only by fight and fection, but in that by the greateft force
made ufe of in compreffing and fquee-
zing it between ones Fingers, we find
it impoffible to force out the leaft
appearance of Humidity through its
aforefaid Inclofure or Integuments.

Being therefore very inquifitive
after the true ufe of this part, and
defpai-

despairing of ever attaining to such a
Knowledge without first knowing the
exact Structure thereof, besides all
other means commonly made use of
in all Anatomical Enquiries, I made
use of all sorts of Injections serviceable
to such an end, as of tinged Liquors, Wax, and Mercury, but all
with little, if any, success according
to my expectation, the Wax not pe-
netrating its Texture at all, the tinged Liquors but very superficially,
and the Mercury, (where my chief
Hopes were) always by its weight
(do what I could to the contrary)
either breaking through the sides
of the *Infundibulum*, where it leaves
the Brain, or else falling down in
greater *Globuli* than the extream
narrow Passages were capable of ad-
mitting, and by this means became
altogether useless.

Being compelled therefore for the
present to leave off a little while a
further enquiry into the Structure
of this part, by reason of the great
mist it is involved in, and to gain
a little more Light for our Gui-
dance in searching after Truth,
(which like many other things of
greatest value lyes deep, and is
with

with great difficulty acceſſible) it
may not be amiſs to ſee what
Aſſiſtance can be had, by making
diligent Scrutiny into the Structure
of its Appendix the *Infundibulum*.

The Infun- This is a thin medullary Duct,
dibulum. covered with the *Pia Mater*, deſcending from the internal Concave
Superficies of the Brain, to which, by
reaſon of its wideneſs towards one
end, and narrowneſs towards the
other, in reſemblance to a Tunnel,
as well as by reaſon alſo of the
parity of their Uſes, the Ancients
gave the Name of *Infundibulum*.

In Man it is cloſely inveſted with
the *Pia Mater* at its very entrance
into the Gland, and from that
place hath not any manifeſt Cavity
I could diſcover by blaſt or ſtyle,
but is altogether of a medullary
ſubſtance, contrary to what it is
in Sheep or Calves, in which laſt
Creature, where the Parts are larger, by inſerting a Blow-pipe into
that part of the *Infundibulum*, next to
the Gland, I have ſeen its further
Tract or Paſſage on the upper part
thereof a little puffed up, and a conſiderable

fiderable quantity of Water regurgi-
tate, as though it had lain contain'd
either in fome Pipes or Porulous Subftance of that Gland.

This Difference is not taken notice of by *Vieuffenius*, and therefore
what he fays of this part feems chiefly in this refpect, if not altogether,
applicable to the Structure it hath in
Men.

Thofe two Divifions or Ramifica-
tions of this part the faid Author men- *Vieuffen.*
tions, one forwardly, and the other P. 49.
backwardly, in Sheep, Calves, &c. par. 3.
I have always found correfpondent
to the Defcriptions he there gives of
them; but whether the firft be
protended fo, and terminate after
the manner he there defcribes I
fomewhat fcruple, feeing I have always obferv'd the Extremity of
that part in Brutes, towards the
foremoft part of the Gland, finking
as it were into the very Subftance
thereof, and afterwards becoming
prefently altogether imperceptible,
and in Man the termination thereof
juft after the fame manner, fave only
that in the laft it happens forthwith
upon its approach to the Gland,
with-

without being protended either backwardly or forwardly.

The Ufe of this part is certainly
to convey fome fort of Humidity
from that great concamerated Cavity within the Brain, refulting from
its inward complication of parts, to
the Pituitary Gland, and the office
of it is to receive and carry oft this
tranfmitted Humidity; but as to
how either this Humidity is collected
in the aforefaid Cavity, or how,
when convey'd into the Gland, it is
carried off, we are ftill as much in the
dark as ever.

I know very well there is nothing
more eafie with the Vifionary Philofophers than fuch a Knack as this;
and now I think on't, the great *Wil-* *Will.* p. 45.
lis makes nothing of turning a Vein
into a *Lymphæduct* in the *Glandula*
Pinealis and *Plexus Choroeides,* no
lefs than which does alfo the accu-
rate *Vieuffenius,* in the *Plexus* belong- *Vieuf.*p.110
ing to the fourth Ventricle; but how par. 3.
confonant this is to the rational ftructure or mechanifm of parts, neither
the one or the other have been fo
kind as to explain.

Now

Now, as to the *Plexus* and *Glands*
before mention'd, 'tis evident by what
hath been already difcover'd and accordingly given an account of in the
preceding Pages, they are furnifh'd
with *Lymphæducts*, as proper reductory Veffels ; fo that fo far the Pro-
phecy is vanifh'd.

But as to the remaining Gland, I
am not fo fond of gueffing to fay it
hath any, and confequently all I can
fay is, that as I look upon the *Infundibulum* to be no more than
a large *Lymphæduct* varioufly ra-
mified through the *Glandula Pituitaria*, difcharging its Liquor by thofe
many fmall Branches into the Veins
difperfed through that part to be
reduced after the manner 'tis in all
other Secretory Glands back to the
Blood again.

And that which feems moft to favour this Conjecture, is the extraordinary humidity of this Gland,
efpecially in Brutes, above the reft of
the whole Body, as ferving not only
to export what *Lympha* is feparated
from feveral Arteries difperfed thro'
it, but that alfo which it is charged
with from the Brain it felf.

And

And to this twofold manner or double office of Secretion is owing the two diftinct Subftances it feems to confift of, the one being accommodated to that part of the *Lympha* coming from the Brain, and is therefore whitifh, the other to that feparated immediately out of the Blood, and is therefore reddifh.

Laftly, As to the manner how the *Lympha* paffes down thro' the *Infundibulum* from the Brain to the *Glandula Pituitaria,* I look upon it to be in the form of condenfed Vapours arifing from the Arteries of the *Plexus Choroeides,* emitted thence for the keeping moift and in good order that inward Production of the *Pia Mater,* fpread all over its *Parietes,* which being a membranous dry part of it felf, might otherwife become injurious to that fine medullary part lying under and being contiguous to it ; in which there is a continual motion of Animal Spirits, whofe Tracts, and confequently they themfelves, through any the leaft intemperance of this Membrane, would be in great danger of either fome obftruction or diforder.

G   And

And that this *Lympha* is only the
refult of the aforefaid Vapours, I am
the more readily enclin'd to believe,
becaufe I never faw Water in that
part of any found Brain, nor unfound
neither, where the *Plexus Choroeides*
was firm; and there was no reafonable ground, by the extravafation of
Serum in fome other remote parts of
the Brain, to believe it had its rife
from thence.

CHAP.

C H A P. X.

Of the Glandula Pinealis.

THE Gland call'd *Pinealis*, from its Figure, is about the bigneſs of an ordinary Pea, prefix'd to the two Prominencies call'd *Nates*, here-after to be deſcrib'd, at the end of the third Ventricle, immediately un-der the broad and hinder part of the *Fornix*, (with which nevertheleſs it hath no connexion, as *Vieuſſenius* Vieuſ.p.71 ſaith it hath) and over that part of the *Rima* in the third Ventricle call'd *Anus*.

'Tis joyn'd to the *Nates* by ſeve-ral Fibrous Roots, and becomes a ſup-port to that part of the *Plexus Cho-roeides* there ſituate.

In an hydropical Brain of a ſtru-mous Boy, I have ſeen it ſwelled to a ſize of three times its ordinary mag-nitude, and by reaſon of the abun-dance of ſtagnate gelatinous *Lympha* contain'd in it, perfectly tranſparent.

G 2   Hence

Hence it moſt plainly appears that
this part is a meer Gland, and, by
what was ſaid before conformable to
what hath been obſerv'd in this hy-
dropick Brain, of the Conglobate or
Lymphatick kind, and by conſequence
a very unfit part to be made a Re-
ceptacle for Animal Spirits, as *Vieuſſe*- *Vieuſ.* p.71.
nius makes it, and much more a place
of reſidence for the Soul, according
to *Des Chartes.*

'Tis true, there are two fair medullary Tracts ariſing ſeemingly from
the two Roots of the *Fornix,*
ſtretching length-way upon the *Tha-*
lami Nervorum Opticorum, as far back
as this Gland, (by *Vieuſſenius* called *Vieuſſ.*p.64
Tractatus Medullaris Nervorum opti- par. 3.
corum Thalamis interjectus, as though
it was only one, and accordingly is ſo
delineated by him, *Tab.* 7. GG, but
indeed is two, one on each ſide) about
which place they turn in, and by a
tranſverſe bending kind of a Proceſs
(by the ſame Author call'd *Tractus*
medullaris natibus antepoſitus) unite,
as he hath exactly obſerv'd : And *Willis,* p. 9.
this, doubtleſs, gave occaſion to the col. 1.
Error of *Des Chartes,* as *Willis* tru- par. 1.
ly thought, (whoſe ſublime and moſt
                                    de-

deservedly-admir'd Philosophy had
doubtless been much more useful,
had he convers'd more with Dissecti-
ons, and less with Invisibility) and
Vieuffenius too, (with whom in
the same Mistake doth agree *Mu-*
raltus and *Willis*) for upon a
more heedful inspection (as was
most evident in the Brain afore-
mention'd) it will be found that
no part of the Process aforesaid,
however near it comes to this
Gland, does in any wise become continuous to it.

Mural.
p. 508.
Will. p. 10:
col 1. par. 3

Whart.
p. 141.

Dr. *Wharton* also stumbled upon
these medullary Tracts, placing
them amongst the Nerves themselves, and ascribes the same un-
reasonable use to them as he does
to the Nerves in many other Parts
of the Body, (*viz.*) of separating
a superfluous Humour from the
Cruca Medulla Oblongata, or *Tha-*
lami Nervorum Opticorum, (being
the same Part, and only on the
other side or upper part of the
Brain, under another denomination)
which he supposes to be the *Commune*
Senforium.

It

It hath Arteries and Veins in
common with other Glands, the
Veins ending in the fourth or inward *Sinus* ; as may the *Lym-
phæduéts* too, when they are con-
ſpicuous.

CHAP.

CHAP. XI.

Of the Brain *in general.*

THAT part of this Treatise relating to the Veffels, being difpatch'd, I fhall in the next place proceed to an account of the Brain it felf, under which term are generally comprehended the *Cerebrum* and *Cerebellum,* and *Medulla Oblongata,* which Parts being in many refpects fo different one from another, may juftly challenge a diftinct and orderly defcription.

The *Brain* then, in the firft place, as diftinct from the other two, is that large and almoft fpherical Body which comes firft to fight in the old way of Diffection, filling the greateft part of all that fpace contain'd in the *Cranium,* confifting of two different Subftances (firft taken notice of by *Archangelus Piccolominius)* both in Colour, Confiftence, and Office, the one being more com-

Piccolom.
p. 252.

G 4   pact,

pact, white, medullary, or fibrous,
the other fofter, greyifh, and glandulous.

The utmoft *Malpighius* (by vertue of his Microfcopes) could do,
was to difcover the Cortical part to
confift of Glands of an oval depreffed Figure, and in his Opinion, of the
Conglomerate kind, (but that how
properly, as alfo his calling the
Nerves their Excretory Ducts, I leave
to the Judgment of others) and the
Medullary part to confift of various
Fibres immerged in and having their
original from the aforefaid Glands,
deriving from them a certain Liquor
call'd *Nervous Juice,* concerning the
Exiftence of which, in the ufual
fence 'tis taken in, as a fluid body,
contain'd and running continually
in the Channel of the Nerves, as
Water in Wooden or Leaden Pipes,
for either Nutrition or Cenfation, is
a thing fomewhat improbable, it being not only poffible, but very eafie
to refolve thofe two *Phænomena's,*
the firft from the Blood, and the
other from the Natural Tenfenefs of
Senfible Parts maintain'd by the fupply of a proper Liquor from the
Blood,

*Malp. de
Cereb. Cort.*
p. 78, 81.
par. 3.

Blood, both in their Originals and continued or elongated Productions ; inasmuch as it doth as certainly circulate in them as in any other parts of the Body. And as to the manner how this is done, it will appear very plain and intelligible, after the innate Structure of the Part hath been more accurately enquired into.

The Curious *Lewenhoeck* made a far deeper scrutiny into these two Parts, being very probably assisted by better Glasses, and from what occurr'd to his view, called the cortical part a *pellucid Vitreous Oily Substance*, (the seeming oiliness of which Substance I attribute only to the stagnating of the pure Liquor, growing cold after death of the Creature,) from such a close and regular Position of the *Globuli* swimming therein, as allows the Rays of Light to pass them without refraction, contrary to what they do in the other or medullary part of the Brain, in which they are so dispos'd that the Light cannot pass them in right lines, and consequently being a little distorted, makes them appear

white,

Lewenh.
de Struct.
Cereb. p.37.

white, notwithstanding *Malpighius* *Malpig. de Cereb.* p. 2. on the contrary neither allows the Parts of the Brain to be diaphanous, nor the Animal Spirits to be any thing a-kin to Light.

'Tis true, even by his own con-feffion, that his moft nice and dili-gent Infpections could not free him from many Scruples about what he faw; yet fome things to our pur-pofe were plain enough, as Reticu-lar Bodies of a red colour, which being larger in the Cortical Parts than Medullary, helps to give it that greyifh or *fubrunneous* colour, as he calls it.

Nextly, a tranfparent Vitrous-colour'd Subftance contain'd in moft minute Veffels; whence 'tis plain there are two forts of Liquors in this Cortical Part, one of a red co-lour, or Blood, contain'd in larger Veffels, whofe *Globuli*, which give it its rednefs, either by reafon of their fize or figure, cannot enter thofe fmall Veffels which with the Fluid contained in them conftitute this tranfparent,cineritious,or cortical part of the *Brain*.

The

The other a tranſparent Liquor,
contained in moſt minute Veſſels, as
aforemention'd ; from whence I am
induced to believe this Cortical
part to be only an Aggregate of
different Veſſels, (as alſo I do of
all the reſt of the Parts of the
Body) containing different ſorts of
Fluids.

Of theſe Veſſels ſome contain a
more compound Liquor, commonly
call'd Blood, which whilſt in that
ſtate, by reaſon of the *Globuli*
ſwimming on it, looks red, and by
reaſon of a tubulous Pore of a proper ſize and figure ſo continued to
the Veſſel we call a *Vein*, that it
undergoes a continual quick circulation.

Another ſort of Veſſels there is
which receive and contain a more
ſimple fluid body, of a thin tranſparent nature, which when in ſome
parts of the Body, gives the name
of *Lymphæducts* to the Veſſels that
it runs in ; but when in theſe Veſſels, which are diſcover'd to make
up the great Subſtance of the Brain,
whether Cortical or Medullary, may
be allow'd the name of *Fluidum Ani-
male*.   And

And this laſt ſort of Veſſels I
look upon to be either a certain
Protenſion of an Artery, by its
ſmallneſs render'd capable of hold-
ing ſuch a ſort of Liquor only as
the laſt ſpoken of, or elſe ſuch a
tubulous production of the Artery
as by its Orifice or Pore anſwers to
the figure and ſize of the Fluid it is by
Nature intended to receive.

Upon the ſame exact Enquiry
made by a Microſcope, the medullary part of the Brain appears to
be of the very ſame conſtitutive
parts, ranged only after a ſomewhat
different manner, which makes this
part appear more white, as was
before obſerv'd. But over-and-above
(if it may be allowable to make a
Conjecture) I am enclin'd to think
the Whiteneſs of this part may be
owing in ſome, if not the greateſt
part, to ſuch a narrowneſs of the
Veſſels diſcover'd here, containing
the pellucid Subſtance aforemention'd
as will not entertain any Fluid whatſoever, without its being firſt re-
duc'd into very minute Particles,
or *Leptometry* ſo called : Which laſt
Veſſels I therefore ſuppoſe to be
                                    only

only yet more Capillary Productions of the aforesaid Cortical Veffels, as they are of the red or
Blood-veffels indu'd with such a
Pore as fits them only for the re-
ception of a moft fubtile, fine, foft
Liquor, which I efteem the true
Medullary and Nervous Juice, which
being contained in its proper *Cap-
fula*, and many of them collected
into one *Fafciculus*, at its egrefs out
of the Brain, being there wrapped
up in more thick and ftrong Coverings made of the two outward
Membranes of the Brain, do conftitute that part we call a *Nerve*,
which having all its Integuments or
Membranous Inclofures always kept
turvid and tenfe by its contain'd
Fluid , after a flow and leifurely
manner continually difpenfed from
the Fountain, and by its growing
more taper towards the place of its
termination, by which means it acquires a greater ftreightnefs or narrownefs of its Pores ordinarily call'd
Fibrillæ, it fo falls out that all inward
Impreffions, upon all occafions, are
the more eafily and fpeedily tranfmitted through it.

The

The very fame notion alfo concerning Nutrition (which in the
trueft fence is only an appofition of
Parts nourifhing to Parts pre-exiftent
to be nourifhed) in the reft of the
Parts of the Body, I have thought
reafonable to entertain ever fince,
by affiftance of the Microfcope I
have plainly difcern'd the Veins to
be only continuations of Arteries,
and the Blood to run in the fame
Channel varioufly modified, without
the leaft fufpicion of Extravafation,
(*viz.*) a continual tranfmiffion of
Nutritive Juice out of the Pores of
Arteries, after many windings like
Tindrils of Vines (*Analogus* to which
the red Reticular Bodies of *Lewenhoeck* feem to be in the Brain,)
grown very capillary into certain
Tubuli's or Pores of a correfponding
bignefs and figure, making up the
whole flefhy part of the Body, whofe
Subftance, when 'tis freed by wafh-
ing or injection of Water, we fee
to confift only of large and fmall
Blood-veffels and Fibres; which laft,
whether Nervous or Membranous,
or fuch as relate to Mufcular Motion, commonly called *Carnous*, I
fup-

suppose to be full of minute diſtinct
Veſſels for the communicating and
receiving their proper Liquors or
Fluids after the manner already ex-
preſs'd, which as contain'd in the
ſaid *Tubuli* or Pores, whilſt they retain their Natural Conſtitution and
Proportion, I preſume it is which
keeps the Habit of the Body plump
and vigorous, the more thin and
languid being perpetually carried
back by the Lymphatick Veſſels,
and a great part wholly extermina-
ted by meer ſimple Tranſpiration;
which I adventure to think is not
only ſuperficial from the Sudorifick
Glands in the Skin, but alſo through
the whole Subſtance of inward parts,
through ſmall *Canaliculi's* or *Mea-
tus's* in even the *Viſcera* themſelves;
by which, not unlikely, we may
gueſs at the Meaning of *Hippocrates*,
when he ſaid, *All things were conſpi-
rable and tranſpirable*.

The minuteneſs of Veſſels is that
which hath ſo embroil'd the Thoughts
of Naturaliſts upon this Subject, and
ſet Realities ſo remote from the Un-
derſtanding, otherwiſe 'tis no Para-
dox to affirm the Exiſtence of *Vaſa
Vaſo.*

Vaforum almoft to *Infinitum*, fome
containing Liquids in a continual
more nimble circulation, others in
a gentle protrufion only : Which
will appear altogether unfurprizing,
if it be confider'd that the afore-
mention'd Ingenious Author hath
computed, that even the 64th part *Lewenh.*
of a Miriad (*i. e.*) of a Ten hundred p. 46.
thoufandth part of any Subftance
but as big as a fmall grain of Sand,
cannot, efpecially if of a rigid or in-
flexible nature, enter thofe little Veffels, which are feen in a retiform
manner diftributed amongft, and fixed to the aforefaid pellucid *Globules,*
which fwimming in thofe little Veffels, are difcover'd to make up both
the Cortical and Medullary part of
the Brain.   As alfo further, that even
the tender Coats of the fmalleft of
thofe Veffels which contain the aforefaid moft minute Globular Fluid Bo- *Lewenh.* ib.
dies, are alfo full of yet far more
minute Veffels than they themfelves
are.

Nay, I am fo far from being fur-
priz'd at this kind of Vafcular Confti-
tution of Parts, that I apprehend not
how Nature could otherwife have
                              acted

acted without the confequence of a
boundlefs Accretion, inafmuch as that
when any parts of a Fluid become
extravafate, they neceffarily lofe much
of their progreffive motion, and if
of a grofs confiftence, are either pro-
fcrib'd by the wider paffages, or of
a finer, through thofe more ftraight
and elaborate (*viz.*) by Tranfpiration;
fo that what Particles of Matter foever continually arrive, for either
the augmentation or reparation of
the Parts, muft (unlefs the ruine
of the Subject do not firft happen,
as we fee it often does in Difeafes
proceeding from fuch Caufes) needs
(if not confin'd in Veffels) advance into a monftrous preternatural accumulation, as being , by
reafon of their grofs confiftence,
altogether uncapable of being carried
off proportionably to the meafure of
their aggeftion, in the form of fubtile
Steams or Exhalations.

Befides a rational explication of the
natural Functions which this Hypothefis furnifheth us with it alfo, feems
to clear a great many Difficulties
which have hitherto puzzel'd the
moft refined Phyfiologifts relating to

the Animal Faculty, fuch as are *Senfation* and *Mufcular Motion* ; of which
laft here in the next place, the other
being referv'd for the laft Chapter,
which treats of *Senfation* and *Motion*
in general.

CHAP.

CHAP. XII.

Of Muscular Motion.

TO recite the Opinion of others upon this Subject would be a thing altogether useless here, seeing an Abstract of them is already extant in the *Philosophia vetus & nova* by Mr. *Colbert*; and besides, the most correct of them are not only very unprobable, but absolutely repugnant to plain Reason and Matter of Fact too; an Instance whereof you may have in Dr. *Willis*'s Tendinous Reservatories of Animal Spirits, in Dr. *Mayow*'s Twisting or Fiddle-string Fibres, with whom of late Mr. *Regis* agrees, by which the Muscle must needs lose a great deal of its thickness, than which nothing is more contrary to Experiment; in *Duncan*'s first and second Element of *Des Chartes*, which he makes the Animal Spirits to consist of, contrary even to the very Principles of that great Man's Philosophy, which al-

Willis *de Mot. Musc.* p. 35.

Mayow *de mot. musc.* p. 73.

Dunc. p. 90

H 2   lows

lows no Elaſticity to thoſe Bodies
themſelves, though the Authors of
it in all others ; likewiſe in Dr. *Croon's*
making the Blood it ſelf, as well as
the Animal Spirits, to be mov'd by
the power of the Soul to any
Muſcles ; as likewiſe the extravaſa-
tion of thoſe two Liquors firſt into
the ſpaces betwixt the Fibres, and
then their introvaſation into the Fibres themſelves again, in order to
make inflation, an Error incident to
the Immortal *Borellus* alſo, whoſe imaginary Diſcourſe upon this Subject
ſeems of a very different Thread
from the reſt of his Excellent
Works.

Croont, p. 23, 24, 25, 33. Philoſ. Cell. p. 23.

Borel. de mot. Anim. p.ult. prop. 23. & pluribus aliis locis.

If therefore what hath been already
ſaid about the Structure of Parts
be remembred, (*viz.*) That the Medullary Part of the Brain is only a
Contexture of Veſſels ; that its Nervous Propagation or Nerve is alſo
a Compages of Veſſels, formerly
call'd Filaments, much more narrow
than thoſe of the Brain it ſelf; and,
that theſe Nerves produce, or at
leaſt terminate in the Fibres of all
ſorts of ſenſible Parts whatſoever,
though of a different texture, as
well

well as thofe carnous ones of
Mufcles, which laft are tubulous,
'twill not be in the leaft unreafo-
nable to inferr, That thefe Bodies being kept continually turgid
with the contained Fluid, are equally
capable of tranfmitting or receiving
Impreffions of the Object, as if they
were ftretched longitudinally like a
Bow-ftring from each Extremity,
according as *Borellus* hath obferved.

And as to *Mufcular Motion*, allowing only what may directly be in-
ferred from what hath previoufly
been faid, (*viz.*) That the Nervous
and Carnous Fibres are only a con-
geries of Fluids contained in certain
Veffels communicating with each
other, that by reafon of a Plenitude
in the aforefaid Fibres, the whole
Machine is in a conftant *Equili-
brium*, it will neceffarily follow,
upon the common *Poftulatum*, (to
which all Mankind muft be beholden upon all fuch Explications as
thefe to the World's end) *viz.* that
the Senfative or Rational Soul can
command the Animal Spirits (which
I call only a Nervous Fluid) into a

Primus Impetus, or local motion, that a part of that Liquor, whenever a Muscle contract⟨ed⟩ is tranfmitted through the Veffels which contain it from the great Refervatory thereof, the Brain, to its Carnous Fibres, into whofe Veffels, being fo much narrower than thofe of the Nerves, even by vertue of the fame force which moves it from the Brain, that Liquor is driven after a moft rapid manner, (which Effect, to any acquainted with the nature of Fluids and mechanical Laws of Motion by Projection, needs not any demonftration) caufing the Intumefcence or Inflation of the Mufcle, the fame Liquor at the fame time being driven back again with an equal fpeed from the Antagonift Mufcle into the room of the firft, which was tranfmitted from the Brain to the contracted one, in order to maintain the fame Plenitude or (which is the fame thing in the fence of the old Philofophers) to avoid a *Vacuum*. And if any object the widenefs of the Paffage it is to come back by from the ~~reflexed~~ relaxed Mufcle, as an impediment to an

equivalent

equivalent fpeed in that Liquors re-
troceffion, I have to anfwer, that
the Emptinefs being made firft, is a
fufficient recompence for that.

And here I cannot but take notice, that all they who contend for
Animal Spirits, analogous to thofe
we fee produc'd from various Subjects by Fire, as the only adequate
medium for all forts of *Mufcular
Motion*, have been forced to have
recourfe either to certain Tracts or
Interftices betwixt the Filaments of
the Nerves continued from the
Brain, or the Original of the Nerves
through their whole Productions to
the Mufcle, of which fort are the
Cartefians, or elfe to a certain Nervous Juice, for their place of refi-
dence, of which fort are moft of
the Moderns, and particularly *Vieuffenius*, by which Paffages, or out of
which Juice thefe fine invifible things
are either voluntarily, by the command of the Soul, or inadvertently,
from feveral either inward or outward impreffions, tranfmitted, in
order to produce Motion : which
if true, and the only ways of producing *Mufcular Motion*, I beg leave
to ask, how it comes to pafs, by

　　　either

either of thefe ways, that when another perfon bends my Arm, and that
againſt my Will too, the bending
Muſcles of the Arm become as tu-
mid as when voluntarily or inadvertently contracted at any other
time ; which hath been truly obſerv'd
tho' not ſatisfactorily accounted for, by
Dr. *Croone*, or any other I know of.   *Croone, p. 7.*

But how this or any other ſort
of contraction of a Muſcle happens,
does by the other afore-mention'd
Hypotheſis become explicable, without any manner of difficulty at all :
For when the Cauſe of Contraction
is from the Command of the Soul,
the preſſure is firſt from the Fluid
in the Brain, by which all the inter-
jacent or continued Fluid flows towards the Part to be moved, the
ſame proportion of Fluid being at
the ſame inſtant transferred into its
room from the relaxed Muſcle ; and
when the contraction of the Muſcle
is from the above-mention'd external
force bending the Arm againſt my
will, then the Liquor contained in
the relaxed carnous Fibres or *Vaſcula*
is tranſmitted through the whole continuity of Fluids, to that which is
contracted, and all this without being
be-

beholden to the wild Conceits of a dry and moist part of the Nervous Juice, blind Paſſages, inviſible *Tubuli* betwixt the Antagoniſt Muſcles or Valves in the Nerves, by a meer *Æquilibrium* of the Fluids contained in the Veſſels the Parts conſiſt of.

At the ſame time I am not inſenſible of the Solution ſome have given this Inſtance of Involuntary Motion upon another Hypotheſis, (*viz.*) by ſuppoſing an equality of Tenſion or Elaſticity in all the Muſcles of the whole Body ; by which means it falls out, that when any new additional force (though never ſo ſmall) is added to the Fibres of any Muſcle, as in voluntary motion, or the power of Elaſticity in the Antagoniſt Muſcle, overcome by outward force, as in the aforemention'd Inſtance of Involuntary Motion, the other Muſcle then becomes contracted.

Now, that this is one concurrent Cauſe in both ſorts of Inſtances, as being confirm'd by the Experiment of cutting a Muſcle through, either towards the Extreams or in the middle, by which the Fibres, by their

na-

natural Elaſticity, are found to contract either to one or the other, or
to both Extreams, is allow'd to be
true; but to be the only Cauſe, is
altogether as falſe.

For, in the firſt place, as to the
caſe of voluntary Contractions, it is
allow'd to proceed from a tranſmiſſi-
on of Spirits from the Brain into the
carnous Fibres, (that Hypotheſis of
Steno to the contrary having been
convicted long ſince by *Borellus*, in
his Book *De Motu Animalium*) though
not without the concurrence or ſym-
praxis of the natural Elaſticity of the
Fibres belonging to the Muſcle to be
contracted.

So likewiſe, without the tranſ-
miſſion of Animal Spirits from ſome
force or another, I deny even the poſ-
ſibility of that ſtiffneſs or hardneſs
which is eaſily preſerved in all contracted Muſcles, feeling and ſeeming
as though they were indurated and
ſwelled out, as really they are, whether it be in the caſe of voluntaty or
involudtary motion ; in confirmation
of which, I affirm, that though by
the cutting of the carnous Fibres of
any Muſcle through, which way ſoever

ever it be, the contracted part may, and doubtlefs does, grow thicker by the fhortning of its Fibres, yet by that means only it does not become ftiffer and harder, fo as we find Mufcles do when contracted by any natural Caufe, nor is there any neceffity it fhould do fo, according to any Rules of Mechanifm, feeing the Fibres fhortning only by their own elaftick force, when they find the circumambient fpace give way have no neceffity of fubintration of parts, which is always requifite to procure a ftiffnefs or hardnefs to a part altering its dimenfions as Mufcles do, from a longer and thinner to a fhorter and thicker circumference; and upon this it muft needs follow, that in a Mufcle contracted by involuntary force (in which Action the Brain is altogether unconcern'd) that ftiffnefs or hardnefs then perceivable in it, muft needs be owing to the Fluid or Spirits in the antagonift Mufcle, after the manner already explained, tranfmitted to it.

Now,

Now, to define what fort of thing this Animal Fluid (fo called) is, I fee no occafion to frame any other Idea of it than what we ordinarily have of the pureft Liquors, feeing the Nerves are a Subftance which (to the Senfes of either Smell or Tafte dif-covers very little elfe than what is infipid) are always reckon'd amongft the leaft hot parts of the Body, and doubtlefs far lefs warm in Fifhes than us, who yet have as great a ftock of Animal Spirits as any other Crea-tures. And this Confideration may be it was that occafion'd an Author to give the Animal Spirits the Epi-thite of *Frigidiufculi.*   Du Ham.

'Tis plain enough, that the Veffels T. 1. p. 753 which contains this Fluid are extream minute, and confequently the Con-tent muft needs be of a very fine and depurate confiftence, though without much refemblance to either the aforefaid nimble, faline, or ful-phurous Productions of the Fire.

'Tis in a continual, gentle, direct motion, though perhaps contained in curved or reticulated Veffels, from its original fource to the ends of the carnous Fibres, from whence

it

it is convey'd into the Membranous
or Tendinous Productions, according
as the Fibres terminate, and it may
be by filtration only ; in which, as in
other, and particularly in Glandulous
Parts not subservient to Muscular
Motion, where Nervous Ramifications are very copious, whether it be
of any other use than to keep the
Parts in their proper tone, in order
to their regular discharge of the office
of Secretion, must still remain a Controversie, notwithstanding all that
hath been yet advanced against it,
inasmuch as wastings and numbnesses
of Parts, the common Symptoms of
obstructed or divided Nerves, (which
doubtless by their ~~hastening~~ through *happ*
such Causes to Muscular Parts, gave
the first rise to that Conjecture about
the Existence and Use of that Juice
throughout the whole Body) are
equally explicable by the want of
Tone, as of that supposed Liquor.

To the proof of all this an Experiment frequently made does not
a little contribute, and that is the
injecting the Arteries of a Dog, or
any such Creature, when dead,
upon which there immediately happens

pens a contraction of the Mufcles, according to the different ftrength of them, (*viz.*) of the Extenders in the hinder Legs, and of the Benders in the fore Legs, though the Injection be only of cold Water, the reafon of which effect in particular, if it be remembred what hath been before obferved, (*viz.*) that the Blood-veffels do moft certainly enter the compofition of the Nerves themfelves, will not only become very eafily explicable, but the whole Hypothefis at leaft very highly probable.

If it be faid, That this fpeedy inftantaneous reflux of the Animal Fluid is oppofed by the aforementioned conftant direct motion it hath from its Source to the parts to be moved, 'tis eafie to reply, That its flow direct motion that way is eafily, overcome and repelled by the violent impulfe of the forcibly-relaxed Mufcle the other way.

If further it be demanded, by what means it fo happens that in the Inftance before us of an Arm bent by force, that the refluent Animal

mal Fluid is rather towards the
Mufcle, which by that means then
proves contracted, than towards any
other whatfoever, to all which it
may indifferently have accefs, I think
the Solution feems not difficult, if it
be confider'd, that at the fame time
that the one Mufcle is forced from,
the other is forced into a contraction ; from whence it fo falls out, that
the carnous tubulous Fibres of the
laft, which by being extended under the ftate of relaxation, did lofe
their cavity, muft needs by their
natural elafticity, when freed from
the preponderant force of its Antagonift, acquire it again, by which
means a fpace being made, the repelled Fluid, by the Laws of Libra-
tion, (to fay nothing of the habitual
motion of the Animal Spirits, or
Liquor, by moft Authors, efpecially
Borellus, urged as a Reafon for this
effect) muft needs be driven thither.

In fine, though I am not averfe
to think moft of the *Phænomena*
relating to Senfation and Motion
may be folved by this Theory, tho'
of fo fmall an *apparatus*, yet I am fo
far

far from being fond of it, that I have reſerved a far greater ſhare of Friend-ſhip for any other that may ſeem but of never ſo little more a kin to Truth; and ſubmitting all I have ſaid on this Subject to the candid Sentiments of the more judicious Proceedee in de-ſcribing the other parts of the Brain as they offer themſelves in the uſual modern way of Diſſection.

CHAP.

CHAP. XIII.

Of the Brain *in particular.*

THIS Part being already de-
scrib'd and consider'd in general, as consisting of two different
Substances commonly called its *Si-
milar* Parts, and the Source of all
Sense and Motion, comes now to be
taken notice of in a more particular
manner, with respect to its dissimi-
lar parts or conformation; and this I
think may best be done first according to its outward, and next to its
inward appearance.

Outwardly 'tis convex and cortical, exactly divided into two Hemi-
spheres by the first Process of the
Dura Mater called *Falx*, from the
bony Process called *Crista Galli* forwardly to the very hindermost part
of the *Cranium*, where these two Di-
visions are stretched over the *Cerebellum*, from which part also 'tis perfectly separated by the second Process
of the *Dura Mater*, to the end it

I   may

may not caufe any prejudicial compreffion upon that part, either by its
weight or pulfation.

The foremoft Divifion is made only as deep as the *Corpus Callofum,* the
latter to the very *Medulla Oblongata*
it felf.

'Tis further imperfectly divided
into four Lobes, two whereof (which
being the lefs) are forwardly, and two
(which are much bigger)backwardly.

Thefe Divifions appear beft in the
inverted or *Varolian* Diffection, being
marked out as it were by four Branches of the Carotid Artery, two before, and one on each fide.

Thefe I call *Imperfect Divifions of
the Brain,* becaufe though the *'Pia
Mater* runs betwixt them, together
with the aforefaid Branches of the
great Artery, yet they adhere by
feveral Fibres, both of that Membrane and the Blood-veffels themfelves.

'Tis alfo imperfectly divided thro'
all its external cortical part by the
Pia Mater, though not fo profound-
ly, to the end the Blood-veffels may
penetrate this part in more fine and
reticular Ramifications ; and that by
the

the pulfation of the Arteries the in-
terjacent cortical Glands, (or rather
Veffels) may more freely make their
proper Secretions.

Nextly, it may be confider'd in
its inward appearance, which is concave and medullary, taking its original from the Extremities or *Apices*
of the *Medulla Oblongata;* (or rather
a little more forwardly from the foremoft part of *Vieuffenius's* oval Center)
commonly called *Proceffus Lentiformes,* or according to Dr. *Willis,
Corpora Striata.*

From hence 'tis prefently reflected
back on each fide in the form of a
Vault, very near as far as the
Nates and *Teftes,* a little below which
on each fide 'tis joyn'd with the *Crura
Medulla Oblongata* on their under
fide, being continuous there to thofe
Parts commonly call'd the *Crura Fornicis.*

The middle and uppermoft part
of this Medullary Subftance, by the
Ancients always called *Corpus Callofum,* is therefore by *Vieuffenius* called *Fornix Vera,* in his Opinion
fuftaining that Office (though I fee
not that it does, or for the Reafons

Vieuf. p 61 par. 1.

I 2 before

before given in the defcription of the
Dura Mater and its Procefles, needs
to do any fuch thing.)

This is that part which, as was
before noted, was thought (but miftakenly) by *Vefalius* and others to
efcape the covering of the *Pia Mater*
and in it are not vifible any bloody
Specks, as in moft other parts of the
Medulla Cerebri.

'Tis the *medium* uniting the medullary part of each Hemifphere or
Divifion of the Brain, famous for the
tranfverfe *Stria* running through it
from each fide of the aforefaid
Hemifpheres, the *Septum Lucidum*
only coming between.

In this large or principal Cavity
are contained the three *Ventricles,*
the *Fornix,* the *Septum Lucidum,*
Corpora Striata, Thalami Nervorum
Opticorum, the *Roots of the Fornix,*
the *Tractus Intermedius* of the *Corpora*
Striata, the *Tractus Medullaris Tha-*
lamis Nervorum Opticorum Interjectus,
(which laft has bin already defcribed)
the *Vulva, Anus,* and *Rima* or Paflage
to the *Glandula Pituitaria* by the *In-*
fundibulum , and *Glandula Pinealis,*
(which alfo hath already been defcribed) of all which briefly in their order.

The

IG.5. AA   The three Ventricles, by cutting
afunder the *Fornix* near to its Roots,
and turning it backwards over the
be three Nates, *Teſtes*, and *Glandula Pinealis*,
entricles. appear to be but one, thofe on each
side it being called the *Laterales*, in
which are the *Corpora Striata Tha-*
lami Nervorum Opticorum and *Crura*
Medulla Oblongata, that *Rima*, fo far
as 'tis covered with the *Fornix* and
parts the *Crura Medulla Oblongata*,
being the third.

From the extream Limits of thefe *Vieuff.*T.10
two fide Ventricles, from before to AA, &c.
ntrum behind, does arife that medullary
vale. fpace called by *Vieuſſenius, Centrum*
Ovale, in his Opinion the great
Difpenfatory of Animal Spirits, the
fore part whereof *Willis* calls *Limbus* *Will. de An.*
anterior corporis ſtriati.   *Brut.* p.42.
                                              T. 8. E.

Fornix.   The *Fornix* is a medullary part
IG.5. AA, arifing from two Roots in the fore-
). moft part of the *Baſis* of the *Brain*,
lying betwixt and upon the uppermoft parts of the *Thalami Nervo-*
rum Opticorum, which Roots come
out of the foremoft part of the
Geminum Centrum femicirculare, fo
çalled by *Vieuſſenius*, like two large
I 3   Nerves,

Nerves, and afterwards joyn together, conftituting a broadifh medullary Body, which after having firft
projected it felf for fome fpace forwardly betwixt the *Corpora Striata*,
and afterwards run the length of the
third Ventricle, growing all the way
broader and broader, and towards its
edges (by *Vieuffenius* called *Fimbræ*) *Vieuffen.*
thinner; and being reflected backward Tab. 6. D
towards the hinder part of the lateral
Ventricles, like two Arms, commonly
called *Crura Fornicis*, the beginnings
whereof on each fide are by *Auran-* *Aurant.*
tius called *Hippocampi* and *Bombyces*, *Axat. Obf.*
(from whence, I know, he had P. 45.
chiefly obferv'd this part in Brutes, in
which, by vertue of the hinder part
of the *Fornix*, in that place growing
fomewhat thicker, and running over
the hinder and upper parts of the *Th.*
Nerv. Opticorum, which are more prominent in them, as in *Sheep, Calves, &c.*
than in Men) it is made to appear on
each fide like the bending Creft of
the *Sea-horfe*, and is in colour much
like the *Silk-worm*, certain minute
Stria's, which *Malpighius* calls *Fi-* *Malp. de*
bræ, crofling them like Rings ob- *Cereb. p. 5*
liquely, contrary to what the fame
Au-

Author's Account is of them, who
fays thofe *Fibræ* or *Striæ* run upon
them otherwife, *viz.* as they do on
the *Septum Lucidum* (*i. e.* longitudinally) and embracing the *Th. Ner. Opt.*
on their upper part on both fides,
but adhering clofe to them as one continued Subftance on their under part,
(in which place they are called, by
Vieuſſenius, Pofteriores veri fornicis *Viuſ.* p 51.
(viz. *Corporis Calloſi*) *Columnæ*) be-
come there continuous with the hinder
part of the *Corpus Calloſum*, where
it winds down upon the fides of
the *Crura Medulla Oblongata*, and *Ibid.*
makes up that undermoft fpace or
cavity of the two fide Ventricles,
by the faid *Aurantius* called *Ventri-
culi Hippocampi* or *Bombycini*, and
Vieuſſenius called the hinder part of
the *Centrum Ovale*, which by that
kind of curved paffage lofes fomething of its oval figure.

To the Sep-
tum Luci- The *Septum Lucidum* fome of the
dum. Moderns think to arife from the
Fornix, thence afcending to the internal Superficies of the *Corpus Calloſum* ; others from this laft defcending down to the *Fornix,* but moft

I 4   likely

likely from this laſt, where towards
its foremoſt part I have always
found it double, (firſt taken notice of
by *Sylvius de le Boe*) and as *Vieuſſe-* Sylv. de le
nius truly ſays, often with Water in Boe *Diſp.*
its duplicature. *Med.* p.19.
par. 1.
'Tis a very thin, medullary, Theſ. 13.
tranſparent Body, intermediate to
the *Corpus Calloſum* and ſubjacent
Fornix, by means whereof the two
lateral Ventricles are in that place
ſeparated one from another.

The Corpo-
ra Striata,
FIG. 5.
I l, &c.
The *Corpora Striata,* or *Proceſſus
Lentiformes,* are two Prominencies
ſituated ſomething higher than, and in
Men a great part of them on each
ſide (though Dr. *Willis* ſays, where
the *Corpora Striata* ends the *Thalami
Nervorum Opticorum* begins, which
is only ſo in Brutes) of the *Thalami
Nervorum Opticorum,* or *Juga Crurum
Medullæ Oblongatæ,* and are ſo called
from the many white Streaks appearing in them, deſcending obliquely to the *Medulla Oblongata,* with
Cineritious Subſtance coming betwixt them when they are cut ho-
rizontally.

They

They run down on each side
the *Thalami Nervorum Opticorum* as
far as till the *Corpus Callosum* begins to wind back upon the *Crura
Medulla Oblongata*, towards the hindermoft part thereof.

I have got them delineated here
exactly true, (tho' by neglect without
the *Striæ*)finding all the Cuts of them
in *Willis* to be from Brutes, except
one, which is done very ill, and thofe
in *Vieuffenius* very falfe, unlefs in Figure the 8th, which alfo wants the
Striæ.

*be Thala-
i Nervo-
im Opti-
orum.*

The *Thalami Nervorum Opticorum*
are two prominent Bodies, more
purely medullary on their outward
Superficies than within, which meeting together like the two topmoft
ftroaks of a Y inverted, conftitute the
uppermoft part only of the *Crura Medulla Oblongata* in that form, the
other or undermoft fide being
quite of another figure; and feeing
they are the immediate continued Productions of the *Medulla Globofa Cerebri,* (which contrary to the old Opinion of *Praxagoras* and *Philotimus,*
afferting the Brain to be only a *Ger-
mination*

mination of the *Dorſal Marrow*, of late reviv'd by *Bartholine*, (if any prece- dency of Parts as to time may be al- low'd) I look upon to be rather the original than the production of the *Medulla Oblongata* and *Spinalis* too) and may more properly be called *Capita* than *Crura* of the *Medulla Oblongata.*

Caſſab. in Aſban. p. 137.

Fig. 5. cc. The Tops or *Juga* do, as already obſerved, encline cloſe, yea, joyn together, as *Vieuſſenius* hath rightly obſerved contrary to *Willis*, (whoſe Figures of that part are utterly falſe) unleſs where the *Rima ad Infundibulum* parts them, leaving like the *Corpora Striata* an obtuſe angle between them.

Betwixt theſe two laſt mention'd Bodies there is a medullary ſpace on each ſide, which in a bending manner encompaſſes the *Thalami* themſelves, and receive the Extremities of the *Striæ* in the *Corpora Striata*, as they deſcend from the aforemention'd *Centrum Ovale*, and is therefore by *Vieuſſenius* called *Geminum Centrum Semicirculare*, by *Willis Limbi Poſteriores Corporum Striatorum*

Vieuſ. p. 67 par. 2. Willis de An. Brat. p. 42. T. 8 H.

The

The reason why they are called *Thalami Nervorum Opticorum*, is from certain Fibres supposed to be in them, arising both from their true medullary Superficies (by *Vieuffenius* call'd a *Medullary Membrane*) and some from within their own Sub-stance, which at last, towards their foremost part meeting together, make up the Bodies of the *Optick* Nerves.

Willis says nothing of these Fibres, though in his Opinion *Galen* did not improperly give them that name. *Vieuffenius* paints them very strong.

As for my part, I never could find any Fibres at all appearing in their external medullary part, those within are very small at best, and scarce discernable.

On the outside of these I have always found and often showed a very fair medullary Tract, here descri- bed, running all-along betwixt the *Corpora Striata*,& from the very hin- dermost extent of the *Corpora Stria- ta* forwardly, down to the very Roots of the *Fornix*, to which they seem to be continuous.

A Medul- lary Tract. Tab.5.mm

With-

The Passage into the Infundibulum   Within this Cavity of the Brain are likewise two paſſages into the *Infundibulum*, and ſo on to the *Glandula Pituitaria*, the foremoſt of which is called by the odd Name *The* Vulva. of *Vulva*, and the hindermoſt of *The* Anus. *Anus*, from their ſituation, which with the *Rima* betwixt them, is called, as was before noted, the *third Ventricle*.

The places whence all this Water iſſues are commonly by the latter Anatomiſts deſcribed under the name Tria Fo- of *Tria Foramina*, ſituated ſo as to ramina. give paſſage from all the eminent Regions of the Brain, from whence there can be acceſs had to them for the Water (or rather the *Lympha*, properly ſo called) to fall into the aforeſaid *Infundibulum*, the firſt whereof is behind the *Teſtes*, under the *Valvula major*, (hereafter to be deſcribed) the other juſt under the *Pineal Gland*, or the beginning of the *Rima*, which two meet in an Aperture, under the *Nates* and *Teſtes*, by *Vieuſſenius* call'd *Aquæ Emiſſarium*, having Vieuſ.p.73. par. 3. a ſteep deſcent into the *Infundibulum*; and the laſt at the end of the *Rima*,

Rima, or juft under the Roots of the *Fornix*, and all ending at length (tho' by two different paffages) in the *Infundibulum*.

The Nates and Teftes Tab. 7. CC It may not be unfeafonable in the next place to take notice of two remarkable very fair Proceffes, called **Nates** and *Teftes*, by former Anatomifts fo named from the refemblance they had to thofe parts; but it is plain from thence they were only ufed to diffect Brutes, in which they have fuch a proportion as is betwixt them; whereas in Men 'tis plain they are very near of the fame fize, and not very different in form, being oblong and accuminated towards their Extremities; but in *Sheep*, *Calves*, and moft other Creatures the *Nates* are round and large, and the *Teftes* oblong, fomewhat accuminated, and very fmall.

Before thefe Natiform Proceffes, under the *Glandula Pinealis*, runs a tranfverfe Procefs before taken notice of Pag. 84, by *Vieuffenius* called *Vieuffen. Tab. 8. f* *Proceffus Natibus Antepofitus*, and *Nervuli Æmulus*, which upon further enquiry, by drawing the *Thalami*

mi Nervorum Opticorum ſtill wider,
appears to be rather *Nervi* than *Nervuli Æmulus,* being as thick as that
behind the Roots of the *Fornix*, to
which in ſituation 'tis juſt oppoſite,
and ſeems to joyn the *Thalami Nervorum Opticorum* together, as that
does the *Corpora Striata.*

In what rank to place them 'tis
hard to ſay, as being neither proper
Appendices to either the Brain or *Cerebellum*, properly ſo called, and
being divided from the *Medulla
Oblongata* in ſome meaſure by an In-
terſtice commonly called *Ductus ad
Infundibulum* by the Moderns, but by
the Ancients a Paſſage for the Animal
Spirits to the fourth or noble Ventricle.

The Iſ-
thmus.
They are ſituated upon that part
of the *Medulla Oblongata* which is between the *Cerebrum* and *Cerebellum*,
which ſpace was before called *Iſthmus*,
oppoſite to that part called from its
Author *Pons Varolii*, and by many
Authors, as *Bartholine*, *Spigelius*,
Highmore, &c. thought to be the
two hindermoſt Roots of the *Spinalis Medulla*, which much more likely
Riolanus makes the Proceſſes of the
Ce-

Cerebellum to be, and with him *Vefalius* the great *Vefalius*, who paints them *p.766,767.* fo. *fig.10. AA, I, K. & fig.11. GG.*

From this intermediate fituation Dr. *Willis* thought fit to make them as it were an *Intelligence Office* betwixt the *Cerebrum* and *Cerebellum,* how rightly, I refer to the Judgment of others.

'Tis certain they are medullary Bodies, and contribute to the making the Animal Fluid or Spirits fo called after the fame manner as the reft of the Brain does ; for in cutting them through, (after having taken the reticular expanfion of Bloodveffels off from them, which is very large here, and eminently confpicuous in injected Brains) I find them of the very fame fubftance with the *Proceffus Annularis* and the *Thalami Nervorum optici,* partly cineritious, and partly medullary, and in frefh Brains fomewhat, but very faintly, ftriated.

I know not of any part within the Brain, properly fo called, that is not already defcribed, except a certain *Medullary Chord* at the end of the third Ventricle, and the *Valvula major.* The

Commis-
sura Craf-
sioris Ner-
vi Æmula
of Vieuffe-
nius.

The firft of thefe is a Medullary *Willis* p.43
Procefs, which joyns the *Corpora* col. 2.
Striata together, according to Dr. *Vieuff.*p.83.
Willis, by *Vieuffenius* called *Comiffu-*
ra Craffioris Nervi æmula ; and according to him is the *Medium* or
Commiffura by which his *Geminum*
centrum femicirculare intervening between the two *Corpora ſtriata ſupe-*
riora anteriora & poſteriora, and his
Tractus medullaris tranſverſus & ob-
liquus intervening between his two
Corpora ſtriata inferiora anteriora and
poſteriora, have a communication
with each other.

Dr. *Willis* places this *Chord* or *Willis*, p.6.
Commiffure under the Roots of the col. 1.
Fornix,but it is always behind it, tho'
contiguous to it.

The Valvu-
la major.

The fecond is the *Valvula ma-*
jor, fo called by *Vieuffenius*, but *Vieuf.*p.76.
plainly enough difcovered by Dr. *Wil- Willis*,p.49
lis long before, and its proper ufe col.2.par.2
defcribed.

It is a thick (efpecially in Men)
medullary Membrane, adhering forwardly to the inferiour part of
the *Teſtiforme Proceß*, a little behind that tranſverſe medullary
Procefs from whence the pathe-
rick

tick or fourth Pair of Nerves arise,
laterally to the Procefs afcending
from the *Nates* to the *Cerebellum*,
on its hindermoft Expanfion, to the
foremoft *Vermicular* Procefs of the
Cerebellum, and no where that I know
of to any part of the *Pous Varolii*,
as *Vieuſſenius* will have it, (who
feems to have miftaken another part
for that Procefs) unlefs juft where
the fecond Procefs of the *Cerebellum*
comes out from thence, which jointly
with its fellow Procefs on the other
fide, when they meet together, after
their tranſverſe defcent on the back-
part of the *Medulla oblongata*, do
really make up that part which by
Willis is call'd (and that no doubt
from *Varolius*) *Protuberantia Annularis*, and by others, from its true
Author, *Pons Varolii*.

*Vieuſſen.
p. 76.
Id. p. 73.
par. 3.*

　By raifing up the foremoft above-
mention'd Vermicular Procefs of the
Cerebellum with the Finger, it rarely
fails to come in fight; but if not fo,
'tis eafily fhown, by blowing into
the *Foramen* fituated under the Pineal
Gland.

　Its ufe, according to *Vieuſſenius*,
is to hinder any part of that Water
which falls into the hindermoft *Fo-*

*Vieuſ.p.110
par. 2.*

　　　K　　　　*ramen*

ramen behind the *Teftes*, from running into the fourth Ventricle, or
Vice verfa from the fourth Ventricle
into it, or from getting out on each
fide of the *Medulla oblongata*, over
the afore-mention'd Proceffes, fo as
to fall down upon the Nerves
arifing thereabouts below from the.
Medulla oblongata: Which laft ufe
is evidently moft true, (whether it
be underftood of Water preternaturally or accidentally collected there,
for I muft needs confefs I could never
find any there, any more than I could
in the third Ventricle in Subjects free
from thofe Difeafes incident to that
part, as hath before already been re-
marked *p.*82)but as to that relating to
the paffage from the *Cerebellum* to
the laft or third *Foramen*, I much
doubt the Truth of it, for many Reafons, of which this is one, *viz.*

That the *Plexus Choroeides* in
the fourth Ventricle, together with
the adjacent Parts, being of the fame
Texture as the other are in and about
the two lateral ones of the Brain,
renders it as reafonable to fuppofe
that Water may be collected there as
in other parts of the Brain, (nay, that
it is fo, he himfelf alfo allows as Matter of Fact) and confequently as neceffary

ceffary to have a place of vent for
the Water whenever it happens to
gather there, as it was for that which
was at any time got into the other
Ventricles. And confequently,

In the next place, I do not fee
how this tender Film can be able
to intercept a paffage of fo fearch-
ing a body as Water at any time
forced againft it (notwithftanding the
fuppofed declivity of this Part, which
in Man, by reafon of the largenefs of
the fubjacent prominent annular Procefs, is very inconfiderable) which by
Pulfation muft needs happen whenever
we fuppofe that Cavity filled with it.

And, in the laft place, notwithftanding all the Contrivance the aforefaid Author hath fhewn in conveying
the grofs part of the Water (which, as
was before noted, he grants may be,
nay, conftantly is depofed there from
the Glands of the *Plexus Choroeides*
here fituate) by the Extremities of *Vieuf p. 111*
Veins, out of this Ventricle, I am fufpi-
cious, if there was no fpedier reductory
paffage found out, there would fre-
quently happen very great Mifchiefs
to the *Medulla Spinalis* it felf, and
the Nerves fpringing from it, feeing
the Extremity of that Ventricle called the *Calamus Scriptorius* is there

K 2   par-

parted from the Spinal Marrow behind it, but only by the *Pia Mater*,
which notwithstanding it is there double, as it is also quite down the whole
Spine, left perhaps the Water should
fall down upon the Nerves which arise
from it teo readily, yet upon such an
occasion may be easily suppos'd subject
to violation. Not to say any thing of
the high improbability of any such
Conveyance at all by the Veins, seeing
that in a natural state they are always,
as hath been already obferv'd, continuations only of Arteries.

'Tis true, this may hinder the fall
of Water into the fourth Ventricle, by
reason of a Passage under the *Nates*
before mention'd, by *Vienssenius* call'd
Aquæ Emissarium, so near at hand to
receive it when it finds its further paffage that way obstructed by the in-
terposition and resistance of this
Valve. And for the fame reason
doubtless it was, that in *Vieussenius's*
Experiment which he brings for a
Proof of his Opinion, no Water was *Vieuf. p. 11c*
found in the fourth Ventricle, it ha- *par. 2.*
ving got a passage immediately, upon
its non-admittance by that Valve, to
convey it another way, which by reason of the steepness thereof, is done
much more readily.

CHAP.

CHAP. XIV.

Of the Cerebellum.

THE *Cerebellum* falls next in order to our confideration, in deſcribing of which I hope a great deal of pains may reaſonably be ſpared, ſeeing all that hath been already ſpoken of the cortical or cineritious part of the Brain, as alſo of its medullary part, is equally applicable to the *Cerebellum.* Nor is what hath been ſaid already of the *Plexus Choroeides* in the Ventricles of the one part leſs applicable to that *Plexus* in this.

The Plexus Choroeides *of the* Cerebellum. This *Plexus Choroeides* in the fourth Ventricle begins to be glandulous juſt under the Eighth Pair of Nerves, from whence it runs up on the ſide of the *Caudex Medullaris* to the chordal or third Proceſs of the *Cerebellum,* and from it enters the fourth Ventricle, by *Aurantius* called *Ciſterna*

Aurant. Anatom. Obſ. p. 48.

K 3

Cisterna Spirituum, (which Ventricle, conformably to what that Author hath in the aforesaid place observed, I always find broader than long, and double, though not divided by any intervening Body, as the two lateral ones of the Brain are ;) not lying loose therein, nor at the bottom of it, as the *Plexus* does in the Ventricles of the Brain, but quite contrariwise, (and which hath not heretofore, as I know of, been taken notice of) adhering close to the top of this Ventricle, or the bottom of the superincumbent *Cerebellum,* then running transverse just at the end of the *Calamus Scriptorius,* there becomes continuous to the *Plexus* of the other side ; as hath been obferv'd of the *Plexus* in the lateral Ventricles of the Brain.

This *Plexus* arises from a ramification of the second or backwardest Branch of the Cervical Artery, as one part of the other *Plexus* of the Brain mention'd in that Chapter where the said *Plexus* is treated of, doth) and another smaller Branch of the said Artery about the place where it afcends from the Vertebrals,

FIG. 1. q brals, which laſt Branch turns into a reticular Expanſion firſt, and
then a little ſpace further meeting
with the other, conſtitutes this *Plexus.*

This part differs from the Brain in
its cortical ſtructure, inaſmuch as its
Interſtices are here eliptical or pieces
of imperfect Circles, growing ſhor-
ter towards thoſe two Productions of
the *Cerebellum,* before and behind,
(which by reaſon of certain annular
depreſſions occaſion'd by Bloodveſſels
there embracing them, ſeem as tho'
they were wrinkled like Worms, and
therefore called *Proceſſus Vermicula-
res*) as Parallels upon the Globe do
towards each Pole.

It hath three Proceſſes, which
joyned together on each ſide, make
up as it were two fair Roots, according to the Ancients called the hinder Roots of the *Oblongata Medulla,*
by the Moderns *Peduncles* or *Stalks,*
by which this part grows to the *Medulla Oblongata.*

FIG. 7. gg The firſt of theſe aſcend from
the *Cerebellum* to the *Nates,* the ſe-
FIG. 6. BB. cond from the *Cerebellum* to the
Medulla Oblongata, which meeting
together on the under ſide thereof,

K 4　　　as

as was before noted, make up that large Protuberance by *Willis* called

Var. Anat. p. 26. *Proceſſus Annularis*, by others from the firſt Author *Pons Varolii.*

This I find full of *Stria's* or medullary Tracts, much ſtronger and

FIG. 6: cc larger than thoſe of the *Corpora Striata,* running tranſverſe on each ſide the length of the whole Proceſs, and terminating in a medullary long Tract, dividing that Proceſs into two equal parts, as you

Ib. e e ſee in the Figure, the uſe whereof, as having never been before obſerved,will be hereafter taken notice of.

The third deſcends from this part

Ib. f f backwards, upon the upper ſide of

FIG. 7. hh the *Medulla Oblongata*, like two longiſh thick Chords on each ſide, making the *Medulla* look ſomewhat thicker and broader in that place, and not unfitly ſtiled the *Chordal Proceſs.*

Theſe *Stalks*, when they joyn together at the other end, make up the *Meditallium* or *Corpus Calloſum* of the *Cerebellum.*

The tranſverſe Proceſs of the fourth Ventricle. There are two or three fair medullary Proceſſes cloſe to, and ſometimes ri-

riding one over another, a little on this side the fourth Ventricle, or about the beginning of the *Calamus Scriptorius*, which joyn the two cefles together that defcend from the *Cerebellum* to the *Medulla Oblongata*; and there are two more defeending length-way from that other tranfverfe Procefs behind the *Teftes*, down to thefe.

'ew Pro-
fles on the
fide of the
edulla
oblongata. Thefe long medullary Procefles I never find wanting, though in different numbers, fometimes having feen three, fometimes two, and once I oould find but one, (though larger than ordinary) and conftantly, in what number foever, ending in the tranfverfe Procefles at the afore-mention'd beginning of the fourth Ventricle.

Thefe long defcending Procefles are juft over-againft the *Corpora Pyramidalia*, on the other or under fide of the *Medulla Oblongata*, and the tranfverfe Procefles at the beginning of the fourth Ventricle laft mentioned, are a little above the original of the Eighth Pair of Nerves, infomuch that without being very circumfpect one

may

may miſtake them for the original of
that Nerve, whereas in reality I find
them to be the original of the ſoft or
hindermoſt Branch of the Seventh, as
will be more particularly taken notice
of hereafter, in the deſcription of
thoſe Nerves ; and therefore cannot but wonder how Dr. *Willis* (who

Willis *Ce-* ſpeaks in one place as though he
reb. Anat.
p 12. col. 2. had ſeen them) came to aſſign them
par. 3. for the Root of the ninth Pair,
beneath which and this Proceſs I
have always obſerved the ſpace of
half an inch.

CHAP.

CHAP. XV.

Of the Medulla Oblongata.

THE third part of the Brain, in its general acceptation, according to the foregoing method, is called the *Medulla Oblongata*, all whofe parts on its forefide having already been fpoken of, it remains in the next place that we take notice of it on its other fide, where are most confiderable its *Crura*, fo called, which *Crura* are only the under part of the *Thalami Nervorum Opticorum* before defcribed, which in their Extremities becoming continuous to the under fide of the medullary hinder part of the Brain, occafion'd the Ancients to think the *Medulla Oblongata* had its foremoft Roots immediately from the Brain there, as it had its hindermoft from the Proceffes of the *Cerebellum*; but upon a more diligent enquiry it appears, that thefe

Crura

Crura are more deeply immerged in and knit to the *Medulla Globofa* of the Brain forwardly, by vertue of the *Corpora Striata*, as alfo by the very medullary part of the Brain it felf, which there, from the back or undermoft winding part of the *Corpus Callofum* is perfectly mingled with it.

Where thefe two *Crura* begin to come clofe together, the *Protuberan-* FIG 6. BB *tia Annularis,* or *Pons Varolii,* made up of the fecond Procefs of the *Cerebellum* aforemention'd, begins to cover the *Medulla Oblongata* for about the fpace of an inch and an half, after which this *Medulla Oblongata* in one large Trunk is continued to the firft *Vertebra* of the *Spine,* and fo quite down to the end thereof.

The two white Bodies behind the Infundibulum. FIG. 1. bb Whilft the *Brain* is in this pofition it may not be unfeafonable to take notice of two fair white Bodies on this fide of the *Infundibulum,* in that depreffed part of the Brain, where the *Pia Mater* (as hath before been taken notice of) is fo remarkably double.

There

he Cor-
ra Pyra-
Idalia.
10. 1. n.

There are alſo two white long medullary Proceſſes called *Corpora Pyramidalia* both by *Willis* and *Vieuſſenius,* which ariſe juſt at the ending of the Annular Proceſs running down upon the *Med. Oblongata* the ſpace of an inch, ending a good ſpace below the place where the Eighth Pair of Nerves begin, which have their original between the *Corpora Olivaria* and the *Chordal Proceſſes* partly on the other ſide thereof, contrary to the account we have of them by Dr. *Willis,* who deſcribes them as ending in pointed Extremities, juſt where *thoſe Nerves* have their original.

Willis, p. 13. col. 1. par. 1 p. 61. col. 2. par. 3.

he Cor-
ora Oli-
aria.
bid. o.

On each ſide of theſe appear plainly the *Corpora Olivaria,* ſo called from their Figure, as the former were by *Vieuſſenius ,* which with the *Corpora Pyramidalia* and two white Bodies behind the *Infundibulum,* he calls *Conceptacula Spirituum Animalium,* or places containing Animal Spirits upon ſeveral occaſions of uſe to the Brain, both in its natural and intellectual Faculties.

CHAP.

CHAP. XVI.

Of the Nerves.

IN the fame pofition of the Brain we alfo have a fit time of taking a view of the Nerves, which are ftill medullary Productions of the Brain difperfed to all the parts of the Body, which have need of either Senfe or Motion, and thefe are in number ten Pairs or Conjugations, having their Names and Originals as follows.

The firft is the *Olfactory Pair*, which after they leave the former Lobes of the Brain, and begin to run to the Bone called *Ethmoides*, take the name of *Proceffus Mammillares* ; but this is chiefly in Brutes, where through their largenefs they have that appearance, and are manifeftly hollow.

By the utmoft Scrutiny I have been able to make, they have but one Original, and that is from the undermoft and foremoft part of the

Crura

Crura Medulla Oblongata, where they advance on each side into the Globous medullary part of the Brain, from whence running concealed betwixt its foremost and hinder Lobes obliquely, for a good space, at last they come in sight, as you see them in the Figure: And by what means *Vieuffenius* comes to find such diffused Originals for them as he speaks of, I know not.

Their Use is known to most, and a particular account thereof, as of the rest, together with the manner of Sensation, with relation to the external Organs of Sense, is much more fit for a Physiological Tract than one of this kind.

I shall therefore only at this time give a general description of the Nerves belonging to the Brain, how and where they arise, the difference or variety whereof serve very well to inform us, according to several late Theories, concerning the different Refervatories of the Animal Fluid or Spirits, and the different dispensation of the same to several parts of the Body.

The

The Second Pair.
Ibid. 2 2. The fecond, Pair are called the Optick or Seeing Nerves, of which I find no more Originals than of the former, and that is from thofe medullary parts called *Thalami Nervorum Opticorum*, tho' *Vieuffenius* fays they are from feveral parts; and *Willis* in general terms from the aforefaid *Thalami Nervorum Opticorum*, behind the *Corpora Striata:* which defcription is more exact in *Quadrupeds*, where the *Thalami Nervorum Opticorum* are altogether in fituation behind the *Corpora Striata*, than in Men, where a great part of the *Corpora Striata* are fituated on the outfides of the *Thalami Nervorum Opticorum*, and only their Heads or Extremities before them.

The Blood veffels mention'd both by *Willis* and *Vieuffenius* belonging to thefe Nerves, I have feen to run not only upon or with them, but alfo in injected Bodies exactly quite thro' the medullary fubftance of them, into the reticular Coat of the Eye, wherein they end in an infinite number of the moft capillary Ramifications, which by an injection of that Artery

made

made with Mercury, become very
delightfully confpicuous to the Eye.

The Nervous Fibres alfo, from the
fifth and third Pair of Nerves, do
twine about the Bodies of thefe
Nerves, as the two above-mention'd
Authors do truly affirm, but how
rightly they both affign to them the
office of dilating and contracting
them fubferviently to the vifory faculty, and preternaturally in Con-
vulfions of the Eye, as though thefe
Fibres were truly Mufcles, or of the
carnous kind, I refer to the Judgment
of others.

Thefe go out of the Skull at its
firft *Foramen.*

The Third
Pair.
Fig.1&33
The third Pair arife out of the
forward and upper part of the Annular Procefs, where 'tis contiguous to,
and covered with the under part of
the *Thalami Nervorum Opticorum,*
coming out into fight from between
them, juft where that Procefs terminates forwardly, which is where the
Crura Medulla Oblongata come together into one body, conftituting the
Caudex Medullæ Oblongatæ.

L　　　Thefe

Thefe running through a duplicature of the *Dura Mater*, on the outfide of the *Circular Sinus*, go out of
the fecond hole of the Skull to the
Eyes, and are therefore called *Par
Oculorum Motorium*, to the voluntary
motion of which only they are granted to be fubfervient, which, feeing
they have their original from the *Cerebellum*, afford us no weak Argu-
ment againft the Hypothefis of
Dr. *Willis*, who hath referv'd that
part in Nerves fubfervient to involuntary motions only.

*The Fourth
Pair.
Ibid 44.*   The fourth Pair is very fmall,
coming from the tranfverfe Procefs
on the forefide of the *Medulla Oblongata* behind the *Teftes*, firft coming
in fight between the undermoft part
of the hinder Lobe of the Brain and
the *Cerebellum* laterally, croffing that
part where the Annular Procefs ends
towards the *Crura Medulla Oblongata*,
from whence they pafs into a duplicature of the *Dura Mater*, and afterwards, a little more outwardly than
the former, goes through the fame
fecond hole to the *Trochlear* Mufcle
of the Eye, and are called from their
mo-

moving of that according to the paſ-
ſions of the Mind, the Pathetick Pair.

The Fifth Pair. Ibid. 5 5. The fifth Pair is broad and large,
made up of many thick Fibres conti-
nuous to each other, ſome ſofter than
others, ariſing from the uppermoſt
part of the *Proceſſus Annularis,* which
is backward laterally, where 'tis
broadeſt, by reaſon of the ſecond
Proceſs of the *Cerebellum* there entering it.

The ſeveral Branches of the fifth Pair. This Nerve, after having firſt
climb'd over the inner Proceſs of the
Os Petroſum into a kind of a Cavity
made of a duplicature of the *Dura
Mater* in that place, immediately
ſwells into a kind of a thickneſs, cal-
Fıɢ. 3. B led a *Ganglion,* from whence ſeveral
Branches are propagated, lying betwixt the *Dura Mater* and the *Cranium,* on each ſide the *Sella Turcica,*
without any *Fovea* or Cavity at all,
going out of the Skull at three
Fıɢ. 3. C, D, E ſeveral places, its ſuperiour ſmall
Branch at the ſecond hole with the
third and fourth Pair of Nerves, its
inferiour ſmaller Branch at the third
hole, and its poſteriour or largeſt
Branch at the fifth.

L 2

From

FIG. 2. y.   From the infide of the foremoft
Branch two little ones turn back, and
meeting with another fmall Branch
a little lower turned back alfo from
the fixth Pair, where that Nerve is
faften'd to the outmoft or borrowed
Coat of the Carotid Artery, make up
a fmall Trunk of a reddifh or flefhy
colour, like to that which 'tis of when
paffed out of the *Cranium*, (as *Veflin-*
gius hath truly obferved, who calls it

The Inter-
coftal Pair. *The Internal Branch of the Sixth pair*)
X. z z. which defcending obliquely, and cree-
ping under that Artery, betwixt its
external, proper, and borrowed Coat,
goes out with the Carotid Artery at
the fourth hole of the Skull, which
is in a manner double between the
Os Petrofum and *Cuneiforme*, and from
its paffage through the *Thorax*, near
the Roots of the Ribs, (all-along
which, it receives a Branch from the
Intercoftal Nerves) is call'd , *The*
Intercoftal Pair.

The fixth
Pair.
FIG. 1ß.
6 6.   The fixth are about the bignefs
of the third, arifing from the hinder
part of the Annular Procefs over-
againft, and not far off from the
beginning or head of the *Corpora*
Py-

Pyramidalia. It fends out fometimes one (in this Subject very fhort) fometimes two flips, as was afore faid, for the making up the Trunk of the Intercoftal Nerve, and after that (with the foremoft Branch of the fifth Pair, in one and the fame duplicature of the *Dura Mater*, together with the preceding third and fourth Pair of Nerves) goes out at the fecond hole of the Skull, and terminates in the abductory Mufcles of the Eye.

The feventh Pair. Fig. 1ft, 7 7. The feventh Pair, or Hearing Nerve is large, and comes out almoft juft over-againft the original of the fifth Pair, on the lower or under fide of the fecond Procefs of the *Cerebellum,* where it firft appears coming out from the *Cerebellum* to make the aforefaid *Protuberantia Annularis* between the *Corpus Olivare* and that Protuberance, as though it crept out betwixt them, and had (as it really hath) a more remote extraction.

It confifts of two diftinct Proceffes, the firft of which is more round, hard, and lefs than the fecond, that being for Motion,

this for Senfe, but tho' they feem
as though they had the fame original, being feemingly continuous at
their rife from the Brain, (which
Dr. *Willis* affirms they have, tho'
fometimes he makes it in one place,
and fometimes in another) yet upon
a further enquiry it does appear otherwife, the firft or hardeft having its
original from the *Caudex Medullaris*,
not far from the place where it comes
firft in view ; the fecond very remote
from the tranfverfe Procefs or Proceffes in the paffage to the fourth Ventricle before defcribed, (which in
another place the fame Author feems
plainly to have obferv'd, taking it
for the Original of the other Procefs of this Nerve ;) from whence
it afcends all-along on the fides of the
Medulla Oblongata till it arrives at the
afore-mention'd place, where it firft,
together with the other Branch, leaves
the *Medulla*, to pafs out of it at the
feventh hole in the Bone called *Petrofum*.

Willis,p.12 col.2.par.3 & p. 56. col.1.par.4

Fig. 7.i1

Willis,p.78 col.1-par.1

The eighth Pair.

Fig.1fi88

The eighth, or Par *Vagum*, arifes
a very little beneath the feventh,
but yet not from any part of the
An-

Annular Protuberance, but exactly in that somewhat hollow place betwixt the *Corpus Olivare* and third or Chordal Procefs, having numerous (I have counted ten or twelve) Fibres, but all continuous at their firft rife, for its original.

This in a multitude of Ramifications is fpent upon the Bowels, and goes out at the eighth hole with the Spinal Acceffory Nerve, where the great lateral and the inferiour little *Sinus's* in the Bafis of the Skull go out into the Internal Jugular.

To this eighth Pair about half an inch from its firft rife, whilft it climbs upon or fticks to the *Pia Mater* upon the Bafis of the *Cere-* *bellum,* afcends a Nerve called *Spinalis Accefforius* by *Willis,* but long before him taken notice of, nay, painted and defcribed, by *Vidus Vidius,* the original whereof I find to be as far as the feventh Vertebral Pair, from the foremoft and hindermoft beginnings of that Nerve, notwithftanding *Vieuffenius* confines its

Vidus Vidius, p. 93. T. 18. *Fig.* 2. ✳

L 4   ori-

original to the fourth Pair of that part only.

This Nerve runs under the Vertebral Artery near half an inch on the side of the *Medulla Oblongata*, at length, about half an inch from the beginning of the eighth Pair, leaves the aforesaid *Medulla Oblongata*, running obliquely upon the *Pia Mater* of the *Cerebellum*, to joyn ·with the aforesaid Pair, which it really does in that very place, though it part with it afterwards again.

The ninth Pair.
Ibid. 9 9.

The ninth hath several (in one Body I counted seven or eight) pretty large Fibres for its original, very distant one from another, the first of them coming higher, from the very top of the *Corp.Olivare*; the next, and several others, are much less, a quarter of an inch lower ; and the last much lower yet, about the ending of the *Corpus Olivare*, or beginning of the tenth Pair, with several others between the *Pia Mater* and subjacent *Medulla Oblongata* ; but after all, its Trunk is very little, about the bigness of the Accessory Pair.

Thro'

Thro' the Fibres of this Nerve there runs commonly a ſmall but very viſible Branch of the Vertebral Artery, at its original; as you ſee in the Figure expreſſed by the Letter *k*
16.1. k. on the right ſide, going out at the ninth hole, together with this Nerve and the Vertebral Vein, which Vein Vieuſ.p.163
Vieuſſenius miſtakenly makes to go out at a tenth hole, foraſmuch as that is never found in Nature, neither need be, ſeeing the tenth Pair goes out at the laſt or great *Foramen*, by which the *Medulla Oblongata* paſſes into the *Spine*.

The tenth Pair.

Ibid. 10 10

The tenth Pair, (which had it a double Original from each ſide of the *Spinal Marrow*, (as all the reſt of the Spinal Nerves have) might much more properly be called the *firſt Vertebral*, inaſmuch as that both a great part of its riſe and egreſs is quite out of the bounds of the *Cranium*) ſerving chiefly the Muſcles of the Neck, it begins with three, and ſometimes more, ſmall Fibres lower a great deal, out of the *Medulla Oblongata*, almoſt an inch below the Trunk of the ninth Pair, and is about the ſize thereof.

It

It goes out of the *Cranium* betwixt the firſt and ſecond *Vertebra*
of the Neck, making its paſſage
through the *Dura Mater* from the
Medulla Oblongata, about half an inch
below the place where the ſaid Artery comes in.

The Structure of theſe Nerves is
conſiſtent of many *Fibrilla's* or
Stria's, a certain number whereof
being firſt encloſed in a production
of that delicate inward *Lamina* of
the *Pia Mater* afore deſcribed and
ſpoken of, makes up a *Faſciculus* or
Bundle, and many of theſe collectively the Body of a Nerve.

In theſe *Fibrilla's* or *Stria's* (be-
tubulous and always turgid, as in ſo
many Rivulets ſpringing from the
main Fountain the Brain, and from
thence diſtributed to every reſpe-
ctive part of the Body) is contain'd
the Animal Fluid, by means whereof there is maintain'd a conſtant
intercourſe betwixt it and the Soul,
and reciprocal acts of Friendſhip betwixt one part and another.

This Animal Fluid I look upon
only as a Body conſiſting of very
minute and flexile Particles, contain'd

tain'd in such a space as allows
them a capacity of being agitated
on all sides by vertue of the subtile
matter, or Æthereal *Globuli* they
swim in, by which means they are
render'd capable of pervading the
narrowest Channels of the whole
Machine, provided its Orifice or
Pore be adapt thereto, in contradi-
stinction to those other sort of gros-
ser Particles of Matter, which by
reason of the narrowness and figure
of the space they are to enter, do
approximate so close, as to become
contiguous in all their Superficies,
whereby they become deprived of
their former expansive agitation,
which is always necessary to make
a Body fluid, and like so many small
Filaments orderly disposed, do con-
stitute the Inclosures or Coats of
those Vessels the Fluids are contained in.

This Animal Fluid I conceive to
be in a continual state of Transpi-
ration, proportionable to the measure of its leisurely production, seeing no more necessity of ascribing
any further Uses to it , besides
those afore-mention'd, than I do to
the

the watery Humour of the Eye,
befides its fervice to Vifion, which
is always in a ftate of frefh production, as by the Excellent *Nuck's* *Nuck de*
Experiment is plainly manifeft ; and *Saliv. &*
Duc. Aqu.
yet, by vertue of Tranfpiration, *Oculor.*
fome way or other, though to us *p. 169.*
not vifible, without any inconveniency to that noble Organ.

CHAP.

CHAP. XVII.

Of Senfation *and* Motion *in* general.

THE Nerves thus conftituted, become accommodated for Ufe in relation to their feveral and diftinct Functions, in fome confifting of Sence only, fuch as are thofe appertaining to the particular Senfories, (*viz.*) the Smelling and Seeing Nerves, as alfo the foft Procefs of the Hearing Nerve, fome Branches of the fifth, and it may be of the ninth Pair, for Tafting ; in fhort, all the Nerves belonging to thofe external Senfories, by way of eminency, and in a lefs eminent or general way all the Nerves of the whole Body, which are diftributed to fuch Parts as by reafon of their ftructure are capable of Senfation only, any of which, as furnifh'd with the Nervous Fibrils, but more eminently the *Cuticula*, may properly be call'd

an

an Organ or Senſory of *Feeling* ; in
others of *Motion* chiefly, ſuch as are
all the whole Syſtem of Nerves, (excepting them only afore-mention'd)
ſes, which though in a leſs eminent
manner, are neverthelefs ſenſitive
Nerves alſo : In others of both, in
all reſpects (*viz.*) either in a more
eminent or leſs eminent Senſation,
and Motion too, with relation to the
different Fibres they conſiſt of in
their Originals, as the fifth and ninth
Pairs.

Theſe two different Functions of
Senſation and Motion are executed
after two as different manners.

The firſt of which, being occaſion'd from external Objects, is diſcharged by a preſſure thereof made
on the Inſtrument of Senſe, ſo that
the Motion is backward from one
Extream of the Organ to the other,
where it terminates in the *Commune
Senſorium*, commonly ſo called, and
is therefore ſtiled *Perception, Paſſion,
or Affection.*

The other is diſcharged by ſome
manner of impulſe upon the Organ
from within outwardly, with a ten-
dency either to acquire ſome Good,

or

or avoid some Evil; by which Impulse, when carried on so far, either
in a natural or moral sence, as to terminate in, or to be executed upon
its proper Object; the Object then
may be said to suffer as before in the
other case it might be said to act,
and the perceptive Faculty now to
act as before it might be said to suffer, and this Action is commonly
called *Local Motion.*

For whose sake, seeing 'tis of different kinds, learned Men have thought
fit to organize or divide the Brain
into two distinct Provinces invested
with several Rights and Jurisdictions
abating the Power of the Sensitive
Soul, which before was looked upon
universal over the whole Brain, allowing it only a principal, but no
absolute Empire there: And this they
have done upon no weak or unrea-
sonable grounds, seeing that *Local
Motion* is not only in many respects
performed without its assistance, but
even against its power of resistance;
as in the Pulsation of the Heart, vermicular Motion of the Bowels, and
in a great measure the Act of Respiration.

Now,

Now, that which hath been taken
from the Brain hath been conferr'd
on the *Cerebellum*, to which, though
some Power in this Affair may just-
ly be allowed, as was before observed, yet possibly not altogether so
much as there hath been.

Dr. *Willis*, who is Chief in this
Cause, having distinguish'd *Motion*
into voluntary and involuntary only, hath made the *Cerebrum* accoun-
table for the one, and the *Cerebellum* chiefly for the other; and to
that end hath furnish'd it with the
like number of Nerves, as in his
own words is expressed, *Ut divisum* Willis c.1
cum ipso (i. e.) *Cerebro, imperium
Cerebellum habeat*; nay, considering
the Intercostal Pair, derived from the
fifth and sixth Pair, which belong to
the *Cerebellum*, he hath made it ex-
ceed.

I am apt to think that Learned
Person too soon fell in love with
his first Thoughts, the ordinary
reason of either ones seeing false, or
not far enough.

No-

Nothing being more apparent, than that moſt of thoſe Actions or Animal Motions he calls *Involunta-ry*, and of which he gives ſo many Inſtances, are equally found in Brutes and rational Creatures too, whilſt in the ſtate of Infancy, as well as when grown up, with *this* only difference, that all of them in the laſt are under the con-trouling power of the Soul, and conſequently may be ſuſpended up-on a reflex'd Act of the Underſtan-ding ; whereas in Brutes and Infants they are neceſſary, and do as natu-rally enſue upon the impulſe of the Object, as Water, when unconfin'd, runs towards a Plain.

Now, if all theſe were ſuppoſed to be under the power of the *Cere-bellum* only in Brutes and Infants, the Brain it ſelf muſt neceſſarily be thought altogether uſeleſs in them.

It will be neceſſary therefore to take notice, that there are two ſorts of Animal Motion in Brutes, as in Rational Creatures, the one purely natural, ſuch as is Pulſation of the Heart, and various contraction of the *Viſcera*, proceeding from a certain por-

M   tion

tion of the Animal Fluid continually difpenfed to the Nerves in an
equal proportion, and fo may be
faid to have their caufe origi-
nally co-exiftent with the Creature,
and always prefent : And this kind
we find by a moft convincing Experiment hereafter to be mentioned,
to be from the *Cerebellum*, and ab-
folutely free from the dominion of
the Brain, in its ordinary way of acting or influx.

The other is that of *Inftinct*,
relating to the Senfative Soul, or
an aptitude of the Nervous Structure,
to act according to the Impreffions
made upon the Nerves, either from
within, or from without, and fo may
be faid to depend on the prefence
of fuch Caufes as are fupervenient
and extraneous to Nature, fuitable
to the impreffions whereof the *Animal* either purfues or avoids the
Object, obeys, or refifts the Impulfe.

Now, I take it for granted, that
nobody will deny but that the Nerves
(by vertue whereof thefe laft actions
of *Inftinct* are performed) whether
they arife from the *Cerebrum* or *Cere-*
                                     bellum,

bellum, are equally under the command of the Soul; or elfe, as I faid
before, the Brain in thofe Creatures is
to no purpofe.

And of this fort I reckon all thofe
actions in rational creatures of *Inftinct*
before they have attain'd to the ufe
of their *Underftanding*, from any fort
of Impreffions, or Inadvertent and in-
confulted, when he hath the con-
trouling power of Reafon allow'd him
and makes no ufe of it, fuch as are
called *Habitual*, which at firft were
produced by command of the Rational Part only, but through frequent
repetitions at laft, without any command from that, out of a blind obedi-
ence to a bare impulfe from the Object; or laftly, fuch as happen when
he hath altogether loft the ufe of it,
as in Sleep or Diftraction; in which
laft Cafes 'twill be very difficult to
diftinguifh him from a meer Machine
or *Automaton*.

Now, from what hath been faid,
I cannot but think it plain, that
many of the Actions before fpoken
of in Dr. *Willis*'s fence, by him
called *Involuntary*, as proceeding

from the dominion of the *Cere-bellum* only, fuch as he calls the various Configuration of the Face, from fome Impulfe or Provoca-tions in the *Vifcera* or elfewhere, erecting the Ears, turning the Neck and Eyes about, fudden Shrieks and Outcries upon fome extraordinary frightful Object furprizingly affect-ing one Senfe or another, furnifhed with either fuch Nerves as he fup-pofes to be altogether under the command of the *Cerebellum*, as the fifth and feventh, or elfe to have a very near correfpondence with that part by vertue of Vicinity, as the ninth, do more truly pro-ceed from that perceptive faculty, or (to ufe his own words) that part of the Soul, he hath confin'd to that part of the Medullary Syftem called the *Cerebrum*, inafmuch as in reafo-nable Creatures they may and com-monly are fufpended, as well as the Nerves they flow from, fometimes made ufe of as Inftruments of Voluntary Motion by it alfo ; and to think the contrary, is as much

as

as to fay, that when any body
happens to exprefs any of the afore-
mention'd involuntary Acts, or but
hit his Bedfellow a box of the Ear,
whilft afleep, all thefe muft be allow'd to proceed only from the Organ
of Involuntary Motions called the
Cerebellum.

And of this kind alfo in a great meafure I reckon Refpiration, concerning which I cannot eafily be brought
to think it fatisfactorily explain'd by
Dr. *Willis*, from the Energy of
thofe Animal Spirits which flow
only from the *Cerebellum* in the *Par
Vagum*, after the fame manner they
do to the Heart by the Intercoftal
and that Pair for its pulfation,
and as only under the command of
the Soul, to be ftopt now and then,
as it pleafes, by vertue of fome
Nerves communicated to the Intercoftal Mufcles and Diaphragm,
the chief Inftruments of breathing,
from the *Spina Dorfi.*

M 3   I am

I am therefore rather enclin'd to
think this Motion is of the other
different kind before spoken of, under the Title of *Inftinct*, proceeding from an extraneous fupervenient Caufe, acting conformably
to the courfe of Nature in o-
ther Cafes of the fame kind, as in
Hunger and Thirft, and the like,
where the obtaining the defigned
End or Effect renders the part from
whence comes the Motion for fome
time infenfible of the impreffion,
and where, after the ceafing of the
Effect or Motion, the fenfe of the
impreffion revives again, whence
there happens an equal reciproca-
tion between the Senfe and Fruition,
or Senfe and Motion.

To apply this account of the
manner and reafon of the Spirits
acting upon the Stomach and Pa-
late in relation to Hunger and
Thirft, to that of the *Syftole* and *Diaftole* of the Lungs or Refpiration, 'twill
be needful to take notice, that in an
Infant unborn there is no Refpiration, but yet there is a *Cerebellum* ;
and that if this fort of Motion called

led *Inſtinct*, which I make to differ
from purely *Natural Motions*, ſuch
as are contemporary with even the
firſt living Rudiments of the Indi-
vidual, was altogether and ſolely
owing to the *Cerebellum*, after the
manner of that of the Heart ; then
of neceſſity the Child in the Womb
ought to reſpire. But being ſatisfied of the contrary, it remains that
we account for its reſpiration another way, which is as afore noted,
through the preſence or abſence of
the firſt moving Cauſe or Impulſe,
which I make or ſuppoſe to be any
thing impreſſing the Nerves, propagated through the Organs of
Breathing, ſo as to tranſmit the
impreſſion from within to the perceptive Faculty, preſiding both over
the *Cerebrum* and *Cerebellum* too,
to the end the Spirits may from
thence forthwith be commanded
into ſuch other Nerves as act thoſe
Muſcles which ſerve for enlarging
the whole Cavity of the *Thorax*, in
order to let the Air into the Lungs
more plentifully, which was the
thing aimed at by Nature ; and theſe

M 4　　　are

are the Intercostal Muscles and Diaphragm.

Now 'tis easie to conceive, that whilst the Child is enclosed in its Mothers Belly, there is not that occasion for Respiration as when 'tis born, the main Stream of Blood all that while finding no passage thro' them, and that which does by the *Ruyshian Artery* made of Juices much more mild and cooler, the native heat being little, and the Aliment meer Chyle or Milk; from whence it falls out that the Pulmonick Nerves go altogether unprovoked, which after birth are continually otherwise impressed or provoked by the hot *Effluviums* of Blood, now bred of stronger Food, and by a stronger native heat, and wholly flowing through them; which heat continually, as the Child acquires a greater maturity, encreasing, may, for ought I know, not a little contribute, by way of natural impulse, to its exclusion.

The truth of this will the more clearly appear to any who will take the pains to consider well of the structure of Parts in Children un-

unborn, in whom the ufual circuit
of Blood through the Lungs, which
are defigned tor rarifying and perfecting the mixture of Blood and
Chyle, is denyed; as alfo through
the Liver, ferving chiefly for fepa-
rating that grofs Excrement the
Gall, not bred (at leaft in any
proportion) in an Infant unborn,
and in lieu of thefe, other Paffages,
(which become altogether unnecef-
fary after birth) provided by
Nature after a fhorter and more
compendious way, (*viz*) by the
Foramen Ovale betwixt the *Vena Cava*
and *Vena Pulmon.* and *Tubulus Arterio-
fus* between the *Art. Pulm.* and *Aorta*
in the Lungs, and the *Tubulus Venofus*
between the *Sinus* of the *Porta* and the
Cava in the Liver ; as hath been moft
fagacioufly obferv'd by the late Learned Dr. *Walter Needham.*

'Tis true, That in feveral Creatures there are fome Nerves very
much depending on the *Cerebellum,*
as are they which minifter (though
in a different manner, as hath already been taken notice of, and will
be hereafter further explained) to the
Natural and Vital Functions, (*viz.*)
the

the *Par Vagum* and Intercoftal Pairs, and therefore the aforesaid Author, who is in this as in many other of his Difcoveries very fortunate, and highly commendable, made a very good guefs when he brought thefe Faculties into fubjection to that part, inafmuch as by fe-veral others, as well as by my own Experience upon living Bodies, we find, that notwithftanding moft part of the Brain be pared off with a Razor, yea, even after the *Medulla Oblongata* be divided betwixt the *Cerebrum* and *Cerebellum*, and taken wholly out of the *Cranium*, the Heart will beat, whenat the fame time if the *Cerebellum* it felf be but cut in pieces, though all the reft of the Brain be kept entire, the Creature expires prefently.

Yea, I have feen Refpiration (which only in part depends on the *Cerebellum*) totally to ceafe upon only a fudden violent compreffion of that part by a blow, and, after its being wounded, the Heart to ceafe beating immediately.

All

All which muſt of natural conſequence fall out upon the Hypotheſis, That thoſe Functions of
Nature do depend on the *Cerebellum* for their ſource and influence, which is conſtant, uninter-
rupted, and out of the arbitrary
juriſdiction of the Brain ; yet with
this difference, that in Motions
purely natural, and either contemporary with the *Embrio*, as the firſt
ſigns of its vitality, ſuch as is Pulſation of the Heart, during its encloſure
within the Mother, or ſupervenient
upon its further growth and
more viſible organiſation of Parts,
as the natural contraction of the
other *Viſcera* ſubſervient to the
offices of Protruſion of the Chyle,
ſeparation of the Glandular Juices,
and proſcription of the Excrements,
the Animal Fluid or Spirits do altogether flow from the *Cerebellum*,
the Nerves there both deſcending
from the *Cerebellum*, and terminating in thoſe parts afore-mentioned ;
whereas in Reſpiration, which I call
a Motion of Supervenient Inſtinct,
(if I may be allowed to uſe the word
Inſtinct

Instinct in that fence) the Nerves descending from the *Cerebellum*, and propagated through the Lungs from the *Par Vagum*, ferve only to convey the firft Impulfe or Impreffion of the Object to thofe parts which are by Nature framed and qualified to produce Refpiratory Motion, and thofe are the Nerves of the Spinal Marrow, receiving the impreffion from the *Cerebellum*, feeing that by the aforefaid Experiment it appears plain, that after the whole *Cerebrum* was divided from the *Cerebellum* and *Medulla Oblongata*, the act of Refpiration continued for a confiderable time entire, which Motion is dependent on the Senfative Faculty prefiding in the *Cerebellum*, tranfmitting the firft Impulfe produced by the eighth Pair or *ParVagum* (as before obferv'd) and communicated thence to thofe Spinal Nerves which act the Inter-coftal Mufcles and Diaphragm.

So that all the office of the *Par Vagum*, which is propagated thro' the Lungs, is to convey the Impreffion from thence to the *Cerebellum*, which by vertue of its connexion

nexion with the *Caudex Medullaris*
(from whence the Ancients rightly
thought that part had its hindermoft
Roots from the *Cerebellum*, as before
taken notice of) it is able to tranfmit it further, as the Senfative Faculty prefiding there fhall direct, and
that too by the common way, the
Medulla Oblongata and *Spinal Nerves.*

And further ; That this part is as
capable thereof as the *Cerebrum*, and
is not wholly and only deputed for
the fervice of fuch Nerves or Organs as are employed by the involuntary part or portion of the
Soul, (as Dr. *Willis* would have it)
appears in that the third Pair of
Nerves, by him allowed to be a-
mongft the number of the other
kind of Nerves, (*viz,*) thofe commanded by the Will, from hence
(as hath been already fhewn) hath
its original. And here alfo further-
more give me leave to add , by
way of conjecture, that the reafon
why the Soul hath not an equal
command over thofe afore-mention'd
Nerves dedicated to the vital and
natural Motions, is, the early date

or

or commencement of the office of
thofe Nerves, by which means they
contract an habitual irrefiftible Influx, much lefs fo in thofe belonging
to the Refpiratory Functions, the
exercife whereof is of a later date ;
and laftly,the Influx is not in the leaft
fo habitual in thofe other fubfervient
to the Organical Functions of the
Limbs, inafmuch as they are not
capable of being exercifed till a much
longer time after, and then not fo
uninterruptedly as either the firft
or the fecond, but gradually, and
with intermiffions.

So that the only reafon why upon
cutting the *Cerebellum* Refpiration
ceafes, is, that by that means its
ftructure is difcompofed, and render'd unfit either to receive or
tranfmit the impreffion further to
the aforefaid Nerves, which are fubfervient to the Inftruments of Refpiration.

'Tis true , there are reciprocal
communications betwixt the Nerves
of the *Intercoftal Pair, Vertebræ*, and
Diaphragm, yet feeing they terminate not immediately in the Parts
of

of each others particular diftinct ju-
rifdictions, there is no interchangeable
act or office from thence produced
betwixt them.

For as, notwithftanding there are
fo many Branches of Nerves communicated from the Spinal Nerves
fubfervient to voluntary motion, to
the Intercoftal Pair, on their defcent to
the *Vifcera*, and yet by reafon of their
not terminating in thofe parts, they
are not in the leaft able to bring
thefe Nerves under the commands
of the Rational Soul, by which
provident Care of Nature it fo falls
out, that 'tis not in the power of
any, by mifguided Reafon, to act
injurioufly to themfelves : So
by vertue of feveral Branches re-
ciprocally communicated from the
Intercoftal Pair in its paffage down
to the *Vifcera*, to the Spinal Nerves,
there is no power given to them
of moving the Mufcles to which
they are fubfervient uninterruptedly, after the meer manner of the
Vifcera.

But

But now, to return to where we left off, in some Creatures it's very plain, that Nature hath extended this imperial residence of the Soul beyond the *Cerebellum*, even as far as the *Spinalis Medulla*, having not only put this last motion, but that of Pulsation too, under the jurisdiction of that elongation of the Brain; as appears in the famous Experiment of the Industrious *Caldesi* upon the Tortoise, which after the Head was cut off lived, and carried its Shell about, the space of six Months.

Besides which, 'tis remarkable, (by way of digression) according to another Experiment by the aforesaid Author made upon that Creature, that after even the Heart and all the *Viscera* besides, were taken out, except the Lungs, that Creature (to use his own Expression) was found so to resist Death, as to turn it self from the inverted or supine position it had been placed in, in order to make the Experiment, to its prone or natural one, and to live and move six hours after. From
whence

whence it appears, that *Muscular Mo-tion* is capable of being performed by the Animal Fluid alone, without the concurrence of the Blood, by most Authors constantly hitherto made to go a share therewith in the performance of that action. *Caldesi,* p. 75, 76.

So that we find Nature hath not stinted it self to one place for the Seat of the Sensative Soul, or Reservatories of the Animal Spirits so called, in order to the discharge of the afore-mention'd Functions, no more than it is at a loss about the maintaining them in their Integrity by other ways, when it hath so fallen out that the natural structure of the Organs, destin'd by Nature to that end, have utterly been destroy'd, of which we have many Instances in the *Anatomical History,* those Functions in several Creatures remaining perfect, where after death there have been found neither any *Cerebrum* or *Cerebellum* at all, or at least such as by their constitution was utterly render'd useless to any such end.

N   Of

Of the firft is an Inftance of the
Learned *Wepfer*, in a Child living
fixteen hours after it was born, and
difcharging all the Duties of Nature that one of its age was capable of, and by the by (which all
the patrons of a nutritious Juice by
the Nerves may do well to take no- *Misc.Curiof.*
tice of) of a very ftrong and good   *An.3 p.120*
habit of Body, whofe Brain, after
death, was found to be only an
heap of Watery Bladders or *Hy-
datides*, except a fmall part at the
bottom of the Skull, lying in a
Sinus made in the Wedglike Bone,
where the Pituitary Gland is commonly found confifting only of
three Medullary Bodies, two of
which being each of the bignefs
of a Kidney Bean, and the third
behind them of a Pea only, from
which indeed there did proceed
fome, but very inconfiderable Nerves,
or Nervous Fibrils, but fuch as none
can judge of a due proportion requifite to fatisfie the Exigencies of
the common natural, and vital
Functions.

The

The truth of which is ftill more plain, and without exception, in another Inftance in the *Mifcell. Med. Phyfic. Gallic.* of a Child living five days after it was born, whofe Head had nothing but Water contained within the inclofures of the *Dura* and *Pia Mater*, without the leaft footfteps of any medullary part at all. *Mifc. Med. Phyf. Gall. An. 3. p. 54.*

Parallel to which two laft Inftances, I had one communicated to me by that curious Anatomift and learned Perfon Dr. *Tyfon*, in a Child born alive, with no more Brain in the Skull than what might lye in a Filbird-fhell, the *Medulla Spinalis* being much larger than ordinary, as though part of the abfent Brain had been fqueez'd down thither.

Of the laft (*viz.* where the natural conformation hath been depraved) there is extant an Inftance in two feveral places of the *Mifcell. Curiof.* in a fat Ox, in which while living there were obferv'd but very little figns of any fuch thing, whofe Brain was neverthelefs after death found wholly petrified. *Mifc. Cuf. Obf. 26. & 130. An. I.*

N 2   From

From all thefe 'tis manifeft the Senfative Faculty is able to anfwer its internal or external Impreffions, by one part as well as another, and that the Medullary Syftem of the *Spinalis Medulla* may become as adequate a Senfory, in relation to the aforefaid Functions fometimes, as either, *Cerebrum* or *Cerebellum*.

And as to the power or influence the Soul in general exercifes over the Nerves, howfoever different in their original, feeing we have already obferved what a provident care Nature hath taken for the preferving Creatures from their own violence, in that it hath not only conftituted the chief Fountain from whence the great current of Spirits is derived, for the fervice of the vital and natural parts, by the Eighth and Intercoftal Nerves, which is the *Cerebellum*, fo as to be free from the commands of the Rational Will in its ordinary way of acting, but hath alfo taken care that not any of thofe Branches which have their originals from Trunks, which are under the power of voluntary dictates of the Soul, fhould

should terminate in such Organs by
which those Functions are discharg'd,
(abare communication betweenNerves
of different Provinces not being sufficient to such ends or offices , as hath
been observed in those afore-mention'd
additional subsidiary smaller Streams
of Spirits flowing to the parts con-
secrate to the natural and vital Functions by Branches propagated from
the Spinal Marrow, to the Intercostal Nerve, all the way of its
descent to the lower *Venter.*)

So we may further also remark, that
as there are some manner of Im-
pressions made upon the perceptive
Faculty, after such sort of a manner as
that it even loses its power over
its own Subjects, (*viz.*) the Nerves,
which are subservient to its voluntary commands, as in *Laughing, Sneezing,*
and *libidinous Erections*, the Organs
by which these Actions are produc'd,
being altogether under the power of
those Nerves subservient to the voluntary dictates of the Soul, and
acted after the very same manner as
those of Respiration, as often as proportionable objects present, and (notwithstanding the assertion of Dr. *Wil-*

lis to the contrary, who makes Laughing proper to Man only, and, by the authority of *Ariſtotle*, Sneezing an Affection proper but to few, if any other Creature, beſides Man) might alſo produce the ſame effects in Brutes, provided · their ſtupid Souls were capable of being equally impreſſed by ſuch Objects as are proper for exciting a rational Laughter, as we ſee they are by thoſe producing the aforemention'd *venereous actions,* ſeeing the want of the *Plexus* **Cervicalis,** of the Intercoſtal Nerves, and two or three ſmall Branches propagated from thence to the Nerve of the Diaphragm (which he calls a Diſpoſition peculiar to Man, and conſequently in his opinion the cauſe of that Affection in him) might be in a great meaſure ſuplied not only by that nervous Branch we find propagated from the inferiour *Plexus* of the *Par Vagum* (which Nerve is equally dependent on the *Cerebellum,* as the Intercoſtal) to the third Brachial Nerve, from which the Nerve of the Diaphragm hath one of its originals, but alſo by that other propagated from the *Thoracick Plexus* of the Intercoſtal

Will p 106

coftal Nerve it felf , to the fame aforefaid Brachial Nerve, into which the Nerve of the Diaphragm is inferted.

So, on the contrary, there are fome Impreffions made upon the Soul fometimes , through which it acquires a power over thofe Nerves at other times in no wife fubject to it, and thofe are the impreffions either of great Joy or great Grief, fuitable to which the Vital and Natural Faculties are made either much more or elfe fo much lefs vigorous than ordinary , as even quite to languifh.

How this comes to pafs, according to Dr. *Willis* in favour of his own Hypothefis, and particularly in relation to the firft, (which allows of no Involuntary Motions, but what come ' from the Province of the *Cerebellum*) is explained by fuppofing an undulating or rowling motion of the firft impreffion upon the Brain out of it again, through the Natiform Proceffes into the *Cerebellum*, and from thence by the Annular Procefs into the Intercoftal Pair of Nerves, and fo to the Nerve of the Diaphragm, (and he fhould, to make this way of explication en-

N 4

entire, have taken in alfo all thefe
Vertebral Branches inferted into the
Intercoftal Nerve, in order to the
moving of the Intercoftal Mufcles,
without which that action cannot
be performed) by a correfpondence
between which Nerves and thofe of
the Face, being all of one family, the
aforefaid Gefture of Laughing is performed.

Now, befides the needlefnefs of
bringing the Conceptions or Im-
preftions of the Brain under a neceffity of being executed by the inferiour Province of the *Cerebellum,*
till fuch time as 'tis proved, that
fuch motions of the Spirits, upon
extraordinary occafions, may rationally be granted, without fuppofing
a regular motion of the fame
through fuch fuppofed Paffages lea-
ding from one Part to the other
at all other times, (the allowing
whereof does neceffarily imply a
capacity of the Soul to alter the
courfe of the Spirits influencing the
vital and natural Organs, at leaft in
fome meafure, at its-pleafure, which
is plainly contrary to Experience;)
I fhall hardly look upon that Hypothefis

pothefis to be any more than meerly precarious.

And further, to fhew, that fnch Effects or Alterations of the Vital Organs happening upon violent Paffions of the Mind, are no way owing to fuch a tranfmiffion of the Animal Fluid from the *Cerebrum* to the *Cerebellum*, as the aforefaid Author fuppofeth, I ask, how it fhould come to pafs that in the contrary Paffion of Grief, efpecially when occafion'd by furprizing frightful Accidents, the Heart fhould fo languifh, as fometimes wholly to ceafe beating, feeing in the aforefaid Experiment we find that Motion felffufficient, by vertue of a conftant irradiation or influence of the *Cerebellum* only, and confequently could not be thought fo to languifh upon fuch occafions for want of thofe Spirits it never ftood in need of.

Without therefore being forc'd to have recourfe to that other Hypothefis clogg'd with fo many difficulties, I think the aforefaid cafe may admit of another manner of explication, confiftent with what I have all-along advanc'd upon this Subject

rela-

relating to the true fource of voluntary and involuntary Actions :
if we fuppofe, that from fuch Impreffions upon the Soul as are either extreamly more or lefs wel-
come to it, (in which cafe the Object is faid to act unproportionably
upon the Subject) it may not only
act accordingly, above its ufual irradiation and force over the *Cerebellum*,
and by that means, as fending the
Spirits either more or lefs copi-
oufly to the Vital Organs, particularly the Heart, the neareft way,
(*viz.*) by the *Par Vagum* and Intercoftal Pair, for that time render
them more vigorous, or more languid in their operations, in proportion
to the difference of the Paffions, juft
after the manner it happens in cafes
of Alienation of Mind or Diftraction, where by the Strength of the
Impreffion, or *Idea* upon the Mind,
it drives the Spirits with fuch an
impetus into the Limbs, as makes
them act with a vaft greater force
than what they were wont to do,
even above the refiftance of Chains
or Bars of Iron ; but alfo it may
tranfmit the Spirits more or lefs co-
pioufly

pioufly, to the Vital and Natural Faculties, the other way freed from the
fubfidiary Nerves of the *Spina* afore-
mentioned, to the Intercoftal Pair,
which fends forth ramifications to the
Heart (in Men efpecially) equally
with, if not more plentifully than the
Par Vagum, and from the Vertebral
and Brachial to the Nerve of the Diaphragm and Intercoftal Mufcles, by
which means it fo falls out, upon fuch
impreffions, that the Organs of Refpiration to the *fight*, and that of Pulfation to the *touch*, are very remarkably affected.

By this means I have endeavour'd to reftore the Brain to a
capacity of putting its own Conceptions or Impreffions made upon
it into execution, without being beholden to its neighbour the *Cerebellum*, and that either in relation
to its voluntary, inadvertent, or
involuntary Acts; where, note, I
make a diftinction between Acts involuntary and thofe of inadvertency, inafmuch as thefe laft, though
they are not with, yet they are
not contrary to the actual confent
of the Will, after the manner of
the natural actions of the *Vifcera*,
                                                        fuch

fuch as are out of the power of the
Will to hinder ; befides which, I
look upon no other in Rational
Creatures (in a ftrict fence confider'd) to be involuntary, forafmuch as 'tis a contradiction to fay
a Voluntary Agent does any thing
againft his Rational Will (though
it may be againft his Approbation)
by which he is only diftinguifh'd
from a Brute : Though Dr. *Willis*
hath all-along ufed the word *involuntario* in another fence, con-
founding it with acts of meer Ig-
norance under the term of *Infcie*,
and thofe alfo done only inadvertently , or without confideration,
under the term of *Inconfulto* ; and
doubtlefs upon this notion of Involuntary Motions built his Hypothefis, which makes all thofe Actions
which are perform'd at any time
without the notice of the Intellectual Faculty, notwithftanding at
other times they are altogether under its command, equally depending
on the *Cerebellum* as thofe purely natural, which are always free from the
power of the firft, and alfo abfolute-
ly fubject to the laft.

Thefe

These Actions I have therefore called by the term of *Supervenient Inflinct*, and being the meer Effect of external or internal Impreffions upon Senfative Bodies, as Ecchoes are to those upon fuch as are only natural, are equally competent to Rational and Irrational Creatures, and capable of being exerted by the influence of the very fame Nerves which minifter to the Senfative Faculty, whether it act advertently or inadvertently in the one, or *fpontaneoufly* in the other, (where, by the way, it may not be altogether unworthy of our taking notice, the genuine fence of that word in Actions performed by thofe Creatures, is much nearer a-kin to the term *Inconfulto* than *Involuntario* in Men) without the fuppofed rambling Motions of Impreffions made upon it, (through Paffages only at fome times or upon eztraordinary occafions made ufe of) out of the *Cerebrum* into the *Cerebellum*.

Now, as to the organifation of this Part, made to confift of various Medullary Prominencies, Appendixes, and Trects, by Nature contrived for aud adjufted to the various functions

of

of the Soul, and difpenfation of the
Animal Spirits thro' the whole Syftem
of the Nerves,which firft are confin'd
to, or made to refide in fuch and fuch
places as fo many diftinct apartments,
viz. the *Commune Senforium* in one
place, the Imagination and Judgment
in another, and the Memory in a
third ; of which there is fuch a large
and formal *apparatus* and defcription
(tho' with great difcrepancy of opinion) in *Willis* and *Vieuffenius,* the one
placing the *Commune Senforium* in his
Corpora Striata only, the other in the
fuperiour and *middle Corpora Striata,*
jointly with the *Centrum Ovale ;*
from both whom *Des Cartes* and feveral others, and with much more
fhew of Reafon, particularly *Mal-*
pighius, differ, placing it in the extream limits of the medullary part
of the Brain, where 'tis continuous
with the cineritious circumaffufed
Part; I muft confefs, that as I have
not been able, by the beft enquiry I
could make either into Brains diffect-
ed whilft frefh, or when boiled in
Oyl, to difcover any fuch actual configuration or difpofition of Parts, as
we find fo formally delineated by either

Malpig. *de*
*Cereb.*p.11.
par. 2.

ther of them , but efpecially the
laft.

So neither do I fee any neceffity
thereof, feeing we may mnch more
eafily, and to the felf-fame ends and
advantages, look upon the Soul as
one iuternal principal Senfative Faculty. and the whole medullary part of
the Brain, as confifting of fuch Fibrils
or *Vafcula's* as in fome places more
nearly in others more remotely communicate with the Nerves propagated
thence to all the external Senfories,
one adequate *Commou Senfory*, by
which that principal Faculty both receives all its impreffions, and accordingly, as by fo many gradations of
one and the fame power, executes or
performs thofe different Functions
commonly going under the aforefaid
Names of *The Common Senfe*, or *Sim-
ple Apprehenfion, Imagination, Judgment,* and *Memory.*

And as to the fecond, (*viz.*) the
Medullary Tracts, by which the
Animal Fluid, as by fo many Rivulets, is derived from the great Pond
or Magazine into many Rivers, fur-
nifhing the whole Body therewith, all
I could find by the moft diligent
                                    fearch,

search, were only those which have already in the preceding Sheets been remark'd, of which, in the first place, are those in the *Corpora Striata,* very large and discernable.

Those in the inward or concave Superficies of the *Corpus Callosum* running transversely by the *Septum Lucidum* into the *Fornix,* and from that longitudinally into its hinder Thighs or Pillars formerly called *Bombyces,* over which they run in a wreathed manner, as was before observed, terminating in the back part of the Lateral Ventricles, enclosed in the hinder Limbs of the Brain, which Ventricles at length terminate in, and are continuous to the subjacent fore-part of the *Crura Medulla Oblongata.*

Those in the *Thalami Nervorum Opticorum* running obliquely down to part of the subjacent *Crura* and *Caudex Medullaris.*

Those of the *Nates* and *Testes* running after the same manner, and terminating so too, only something lower.

Those

Thofe in the Annulary Procefs, which forafmuch as they have never before been taken notice of, I have caufed to be engraved in a Figure by themfelves, whofe Medullary Tracts or *Striæ*, furnifhed with Spirits both from the continuous medullary *Caudex*, and Productions of the *Cerebellum* too, of which the Annular Procefs is made, (by means whereof the Nerves appertaining thereto may be rationally fuppofed to be under the influence of both thofe Parts , conformable to what hath all along been afferted ;) are as vifible, being more thick, and of a far harder confiftence, than that of the *Corpora Striata* themfelves, (tho' upon every attempt of cutting that Procefs, they may not appear fo) and moft of them terminating in a middle Medullary Tract, by means whereof there is the fame inconveniency prevented, at leaft in fome meafure, as there is by that *fepimentum* of the *Pia Mater*, continued from the joyning together of the *Crura Medulla Oblongata*, down quite thro' the *Medulla Spinalis*, (*viz.*) that at the fame time the Nerves on one fide

O   may

may, (as *Molinetti*, tho' in another place of the Brain, hath truly obſerved) by any morbid cauſe, be injured, thoſe on the other may eſcape. *Mol.* p. 104

Concerning theſe, ſeeing they ſeem to have a particular aſpect or relation to thoſe Nerves, whoſe originals we find neareſt them, it may not be unreaſonable to think they are particular Conduits, from whence the ſaid Nerves are furniſhed with Animal Fluid, though at the ſame time we muſt allow a very free communication betwixt them all.

And conſequently, we may ſuppoſe the firſt of thoſe to convey Spirits from the globous medullary part of the Brain next to it, by *Vieuſſenius* called the Superiour Part of the *Centrum Ovale*, down to the ſubjacent medullary part of the Brain, to augment thoſe which are produced lower, and particularly for the ſervice of the *Olfactory* and *Viſory Nerves*, which laſt hath more eminently its Supply from the *Thalami Nervorum Opticorum*.

The

The second sort, or the transverse *Striæ's* of the *Corpus Callosum*, to convey an additional Sup-
plement by way of the wreathed
Tracts in the hinder Columns of
the *Fornix*, to the *Crura Medulla
Oblongata*, where they become continuous to the reflex'd part of the
Lateral Ventricles backwardly, for
the service also of the aforesaid
two Pair of Nerves, but more particularly to those arising lower either on the Annular Procefs or *Caudex Medullaris.*

Those of the *Thalami Nervorum
Opticorum* and Natiform Processes,
the first of which lies upon, and is
continuous to the subjacent medullary part of the *Crura Medulla
Oblongata*, the other to the *Caudex Medullaris*, may be supposed
to derive Spirits on the behalf of
those Nerves which spring from
any adjacent parts, whether on this
or the other side of the Annular
Procefs or *Caudex Medullaris.*

O 2   And

And of this fort are the *Optick Nerves*, which are fupplied *immediately* from the firft of thofe Medullary Prominencies, and not unlikely from thofe fair Medullary Tracts afore-mentioned, running from the Root of the *Fornix*, extending themfelves all the way between the *Corpora Striata* and *Thalami Nervorum Opticorum*, in which laft at length they are obliterate. The Third, Fifth, Sixth, and Firft or hard Branch of the Auditory Nerves, *mediately* by continuity of them with the Annular Protuberance, to all which the other or leffer Medullary Prominencies called *Nates*, by vertue of their continuity with the fubjacent parts, may be fuppofed to contribute fomething alfo: and thefe feems to be better provided for than the reft of the Nerves, inafmuch as befides this way of being fupplied from the *Cerebrum*, they have alfo another very vifible, and much larger, from the Second Procefs of the *Cerebellum*, of which the Annular Protuberance

tuberance is made, and this see-
mingly not without a provident
Design of Nature, seeing the Nerves
which are derived thence are much
larger, and have a greater Task
of service layed upon them than
any others of the whole Brain,
as hath also the *Par Vagum*, or eighth
Pair, which therefore, by vertue of
its insertion between the Chordal or
third Process of the *Cerebellum* and
Corpus Olivare (and not according to
Dr. *Willis*, from the points or extremities of the *Corpora Pyramidalia*)
hath a double tribute of Spirits, one
from the *Caudex Medullaris* or *Cerebrum*, the other from the *Cerebel-
ium.*

And to this End or great Service it looks as though this Process
was furnished with such a Texture
as it appears to have, of strong,
large, medullary *Striæ's*, capable of
receiving and containing a Supply
from both Fountains.

Whence

Whence it may not be unfeafonable to remark, That not without
fhew of good Reafon I have all-
along afferted the Propriety of the
Brain to thofe Nerves in part, allowed by **Dr.** *Willis* to be no further affected by any Impreffions
of the Brain, than as firft con-
veyed from it into the Province
of the *Cerebellum*, and confequently to depend immediately on this laft
for influence entirely in order to
convey Animal Spirits to thofe parts
wherein they are inferted.

Upon the *Caudex Medullaris*, on
its under fide contiguous to the
hinder Extremities of the Annular
Procefs, are fituate the *Corpora
Pyramidalia* and *Olivaria*, over-
againft which are the two long
Medullary Tracts lately taken notice of, feeming to come from
the tranfverfe Medullary Procefs
behind the *Teftes*, and terminating
in thofe other tranfverfe Medullary
Procefles before the entrance into
the Fourth Ventricle on the other
fide,

fide, by which there may be con-
veyed a confiderable Portion of the
Animal Fluid to the Pathetick Nerve,
which hath its rife from the firft
tranfverfe Procefs, and to the foft
or fecond Branch of the Auditory
Nerve, which hath its rife from the
fecond on that fide , and alfo
to the Ninth and Tenth Pair on the
other fide.

And to conclude, From all thefe
taken together , with the reft of
the whole medullary part of the
Brain, the *Overplus* of what is not
fpent upon the inmate Nerves of the
Brain may truly be fuppofed to be
promifcuoufly difpenfed to all thofe
other extraneous ones produced from
the elongation of the Brain, call'd the
Spinal Marrow. In which laft there
is this conformation or difpofition of
Parts differing from that of the Brain,
that whereas in that the cineritious
part is external, 'tis here internal ;
and this for very good reafon, and
by a provident contrivance of Nature, feeing that not only the cineritious part of the Brain ferves for
sup-

supplying thofe Nerves which have their original thence, as well as all the reft of the *Spinal Marrow*, and confequently ought to have the largeft fpace and dimenfions poffible, which without this fituation could not have been; but alfo without this contrivance the Nerves of this part muft of neceffity have had their originals from the cineritious part of the aforefaid Marrow, contrary to both the cuftom and convenience of Nature too.

THE

Fig. 1.

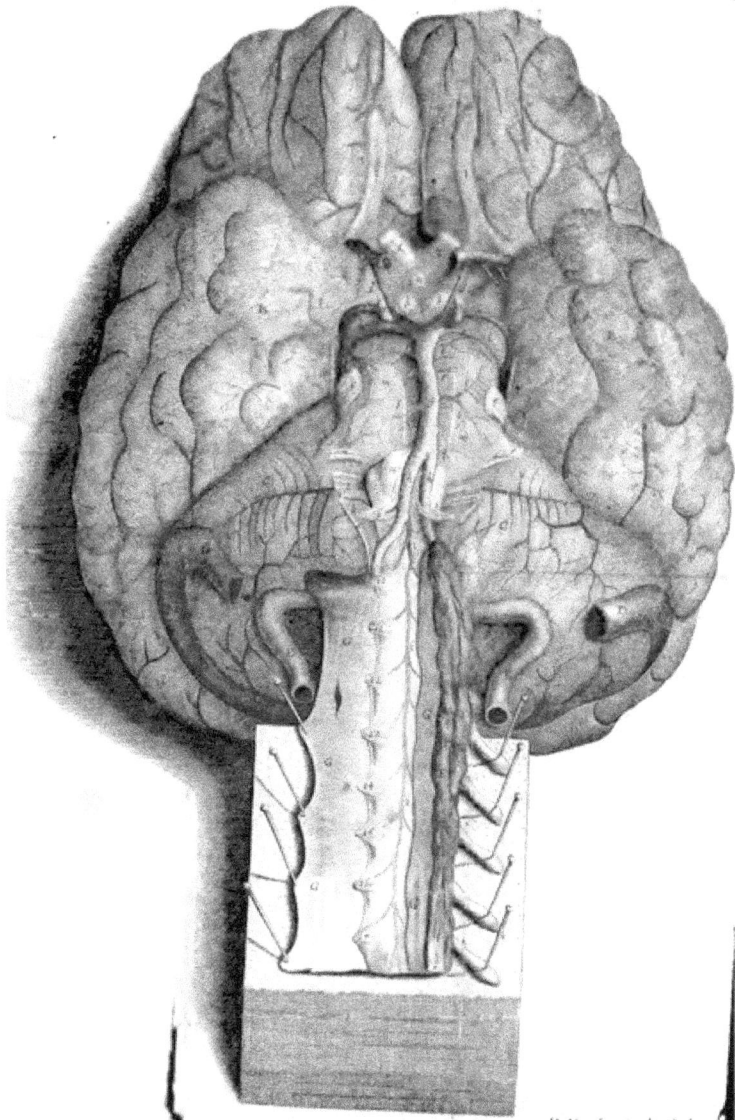

M. Vander Gucht Sculp.

F I G. I.

Exhibits the Basis of the Brain, with part of the Medulla Oblongata, *the Blood-vessels being injected with Wax.*

A A The fore Lobes of the Brain.
B B The hinder Lobes.
C C The *Cerebellum.*
D D The lateral *Sinus's.*
E E The Vertebral Arteries as they pass between the first
      Vertebra and the Bone of the *Occiput.*
F   The Vertebral *Sinus.*
G,&c The *Dura Mater* on the right side taken off from
      the Spinal Marrow, and remaining on the left.
1,2,3 The ten pair of Nerves belonging to the Brain,
4,&c. with seven of the Spinal Marrow.
  a   The *Foramen* that opens into the Pituitary Gland
      from the *Infundibulum.*
b b The two white Protuberances behind the *Infundi-
      bulum.*
c c The two Trunks of the Carotid Artery cut off where
      they begin to run betwixt the fore and hinder
      Lobes of the Brain.
d d The two Arteries joyning the Carotids with the Cer-
      vical Artery, called the Communicant Branches.
e e Two large Branches of the Cervical Artery, some-
      times seeming as tho' they came from the Com-
      municant Branch on each side, from the first of
      which the *Plexus Choroeides* hath its original in
      chief, and from the last the *Plexus Choroeides* of
      the *4th* Ventricle.
f   Several little Branches arising from the Carotid
      Artery.
g   The Cervical Artery composed of the two Trunks
      of the Vertebral Artery within the Cranium.

hh The

The Explanation of the Figure.

h h The two Trunks of the Vertebral Artery.

i i i The Spinal Artery.

k A small Branch of an Artery running through the 9th pair, broken off from its other part thro' inadvertency of the Graver.

l l The *Crura* of the *Medulla Oblongata.*

m m The Annular Protuberance, or *Pons Varolii.*

n That part of the *Caudex Medullaris* on the right side called by *Willis* and *Vieuffenius* Corpora Pyramidalia.

o That part on the same side called *Corpus Olivare.*

p The foremost Branch of the Carotid Artery, dividing the fore Lobes of the Brain, consisting of two Branches, one of them only appearing here.

q q Little Branches of Arteries helping to make the *Plexus choroeides* in the 4th Ventricle.

r r r Branches of Arteries dispersed from the Cervical Artery upon and thro' the Annular Protuberance.

s s Part of the 2d Process, or *Podunculi*, of the *Cerebellum.*

* * The Spinal Accessory Nerve.

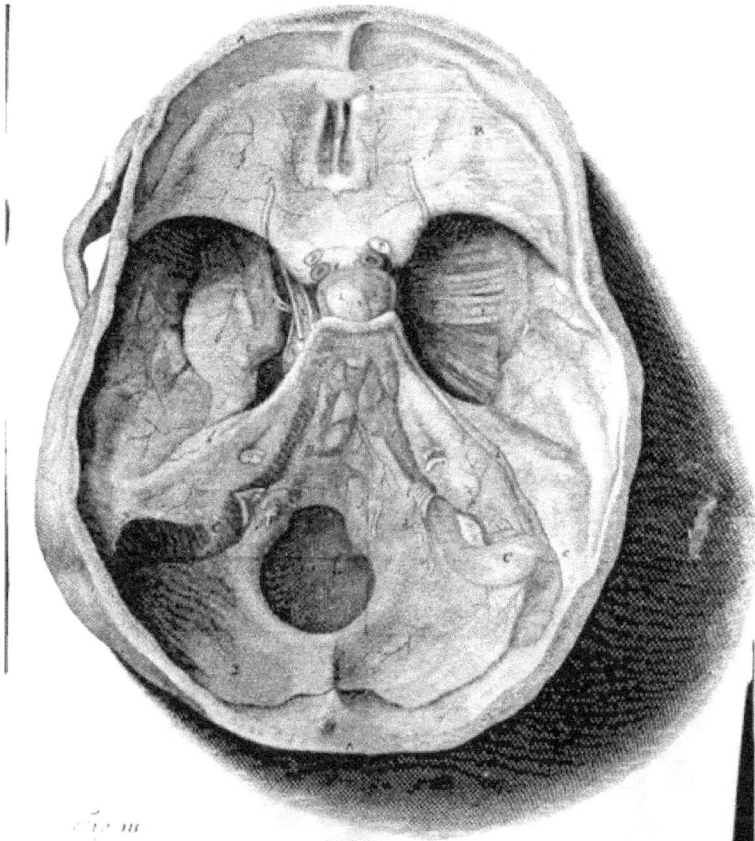

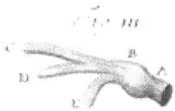

Fig. III

* Vander zuste Tafel.

FIG. II.

Shewing the internal Basis of the Cranium, the Sinus's being injected with Wax.

A A The Edges of the Skull.
B B The *Dura Mater* upon the bottom of the Skull.
C C The lateral *Sinus's*
d d The superiour, longer and narrower *Sinus's.*
e e The inferiour, shorter and wider *Sinus's.*
f The Process of the Bone *Cribriforme*, called *Crista Galli.*
g g Some small descending Branches of Veins upon the bottom of the *Dura Mater.*
h h The first Branch of Arteries proper to the *D. Mater.*
i i The second Branch of Arteries belonging to the *Dura Mater.*
k The third Branch belonging to the *Dura Mater.*
L The last hole of the Skull.
m m Several Veins communicating with the inferiour short *Sinus's.*
n Part of the *Os Jugale.*
o o The *Os Ethmoeid*, where the first pair of Nerves or mammillary Processes go forth.
p p The Optick Nerves cut off.
q q The Carotid Arteries cut off.
r The third pair of Nerves visible only on one side.
S S The fourth pair of Nerves turned up.
t t The fifth pair of Nerves on one side expanded before it is divided into its three Branches, on the other side whole ; which Nerves, with its three Branches, are expressed in the third Figure.
V Its foremost superiour Branch on the left side, going out at the second hole of the Skull.
w The sixth pair of Nerves.

P 2

X The

X   The Intercoftal Nerve, in this fubject proceeding from two Branches of the fifth Nerve, joyning with the body of the fixth Nerve.

y   Two Branches of the fifth pair of Nerves, in this fubject running almoft clofe to the 6th pair, being partly the Roots of the Intercoftal Nerve, which creeps out of the Skull under and between the Coats of the Carotid Artery.

z z The Body of the Carotid Artery, after it has entred the Cranium.

1 1 The *Glandula Pituitaria.*

2 2 The Circular Sinus.

3   The *Infundibulum.*

4 4 The Frontal Arteries.

5   The place where the Lateral Sinus's begin to be declive and tortuous.

6   The *Dura Mater* raifed and reclined to fhew the fubjacent Nerves.

7 7 The feventh or Auditory Nerves.

8 8 The eighth pair, or *Par Vagum.*

9 9 The ninth pair.

F I G. III.

Being the Fifth Nerve, with its Branches, whilft within the Cranium.

A Its Trunk.

B Its Ganglion.

C Its firft or fuperiour Branch, going out at the fecond hole of the Cranium.

D Its fecond or midle Branch, going out at the fecond hole.

E Its third or hindermoft Branch, going out at the fifth hole.

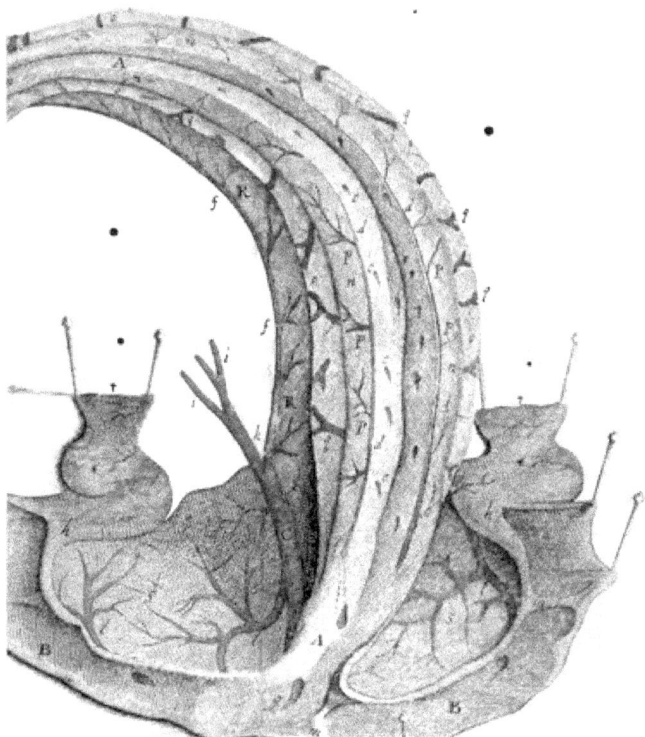

FIG. IV.

*Shews the superiour and lateral Sinus's of the Dura Mater,
opened after they had been injected with Wax.*

A A   The third or longitudinal *Sinus.*

B B   The first and second, or lateral *Sinus's.*

C   The fourth *Sinus.*

d d d   A Vein running on each side of the third *Sinus.*

eeee   Mouths of Veins opening into the longitudinal *Sinus* of the *Dura Mater,* after a contrary manner one to the other.

f f   The fifth *Sinus* at the bottom of the *Falx.*

g   The *Torcular,* where all the superiour and lateral *Sinus's* meet.

h h   The tortuous part of the lateral *Sinus* running under the *Cerebellum.*

i i   The Veins entering the fourth **Sinus** from the P*lexus Choroeides.*

k   The place where the fourth *Sinus* arises.

* *   The *Specus* or round hole at which the lateral **Si-nus's** on each side go out into the internal Jugular Vein.

l l   Two large Veins, whereof one enters the fourth *Sinus* upon the second Procefs of the *Dura Mater,* so as to refift the course of the Blood in that *Sinus,* in its afcent to the Torcular; the other upon the fame Procefs, so as to hinder its defcent to the *Internal Jugular,* contrary to a conformation of Veffels which *Vieuffenius* mentions in his third Table, H H.

mmm   Tranfverfe Chordal Ligaments in the longitudinal and lateral *Sinus's.*

n n   Part of the *Dura Mater* on each side of the longitudinal *Sinus.*

o o   Portions of the *Pia Mater.*

PP&c   Divers fmall Veins on the *Dura Mater,* which enter thofe that run on the fides of the longitudinal *Sinus,* according to its length.

qq &c.

The Explanation of the Figure.

qq&c The Veins of the *Cerebrum* as they appear under the *Pia Mater*, before they enter the longitudinal *Sinus*.

R R The falcated Procefs, with its Veins which enter the fifth *Sinus*.

S S The fecond Procefs of the *Dura Mater*.

† † The beginnings of the Jugular Veins.

F I G. V.

Reprefenting the Brain in a middle fection, the Blood-veffels being firft injected with Wax.

A A The *Fornix* cut off at its Roots and turned back.

b b Its Roots at the beginning of the *Thalami Nervorum Opticorum*.

cc,&c. The *Thalami Nervorum Opticorum*.

d d That part of the *Crura Fornicis* which growing fomewhat thicker as it turns off towards the Lateral Ventricles, runs over the *Crura Medulla Oblongata*, which being very prominent in Sheep, and Calves, helps to thruft it up into fuch a Protuberance as the Ancients called *Bombyces* or *Hyppocampi*.

e e That part of the *Plexus Choroeides* which is made of the firft Branch of the Cervical Artery, fometimes feeming as thô it came from the Communicant Branch, in the Lateral Ventricles.

f The place where thofe two *Plexus's* on each fide meet under the *Fornix*.

g g That other part of the *Plexus* which is made of the fecond Branch of the Cervical Artery joyned with the firft by a Communicant Branch not to be feen here, lying under the *Crura Fornicis*, which is expanded all over the *Ifthmus*, becoming glandulous near to, and efpecially under the *Glandula Pinealis* covered here with the *Fornix*.

h h Two

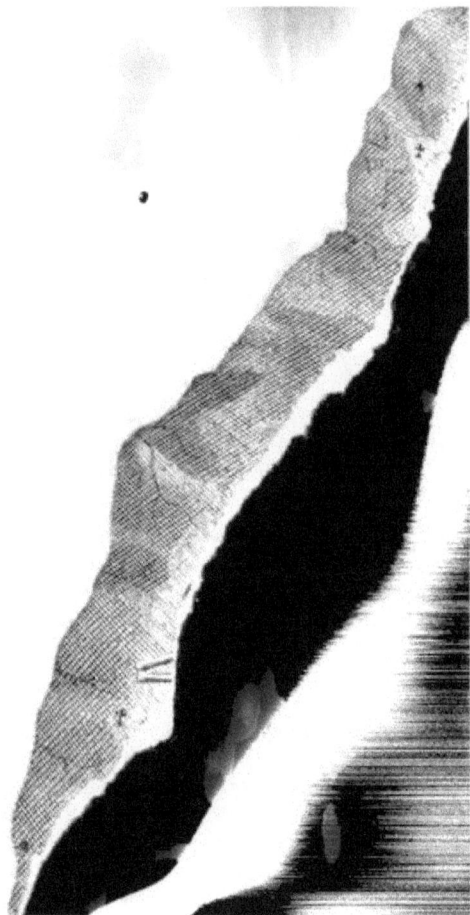

The Explanation of the Figure.

F I G. VI.

Being a draught of the Annular Protuberance, **Med.** Spinalis, &c. *cut through the middle lengthway.*

A A   The *Crura Medulla Oblongata.*
B B   The Annular Procefs, or *Pons Varolii* divided.
c c   The Tranfverfe *Striæ.*
e e   The intervening Medullary Tract in which the
       Striæ terminates on each fide.
f f   The third or chordal Procefs of *Dr. Willis.*
h   The Spinal Marrow.
i i   Some part of the *Cerebellum.*
k k   The fecond Proceffes of the *Cerebellum,* which com-
       pofe the Annular Protuberance.
l l   The cineritious part of the *Medulla Oblongata.*

F I G. VII.

Being the Cerebellum *cut through on its hinder part, and
reclined laterally.*

A A   The *Cerebellum.*
B B   The arboreous ramification of the *Meditallium* of the
       Cerebellum appearing, being cut right downwards.
C C   The Pathetick Nerves.
c c   The *Nates.*
d d   The *Teftes.*
e   The tranfverfe Procefs whence the Pathetick Pair
       have their original.
f   The *Glandula Pinealis.*
g g   The firft Procefs of the *Cerebellum* running from it to
       the *Nates* here extended laterally.
h h   The third or Chordal Proceffes.
i i   The tranfverfe medullary Procefs in the 4 Vent. from
       whence the foft Branch of the 7 N. has it original.
k k   The Medullary Procefs defcending from the Tranf-
       verfe Procefs behind the *Teftes,* down to the afore-
       mention'd other Medullary Tranfverfe Procefs.
l l   The Originals of that Procefs a little too low.
m m The eighth pair of Nerves.
n   The *Calamus Script.* or Extremity of the 4th Ventricle
o   The Spinal Marrow.
P P   The Acceffory Nerves.
q q   The tenth pair of Nerves.

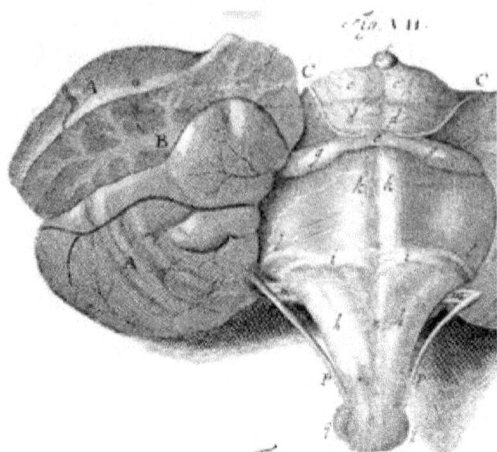

Fig. VII.

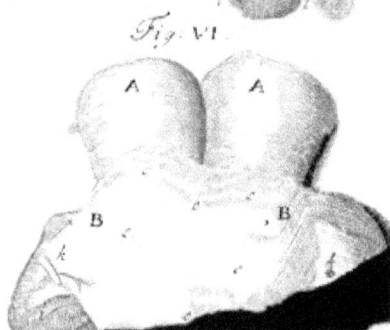

Fig. VI.

THE
TABLE.

                  Q.   The

The TABLE.

How

The TABLE.

Q 2   Its

The TABLE.

                                                         E

How

Q 4 Lym-

The TABLE.

N

N

O

The TABLE.

O

P

The TABLE.

The

The TABLE.

The

The TABLE.

The

The TABLE.

Some

The TABLE.

FINIS.

ERRATA.

PAGE 9. l. 14. for *to* read *towards* ; p. 16. l. ult. for *from which* r. *which from* ; p. 32. in the title of the Chapter, for *Veins* r. *Vessels* ; p. 32. l. 13. after *Veins* insert *which last have already been treated of* ; p. 64. l. 5. dele *only* ; p. 89. l. 16. *Vitrious* r. *Vitrous* ; p. 92. l. 29. for *Septometry* r. *Leptometry* ; p. 102. l. 3. for *contracted* r. *contracts* ; Ibid. l. 29. for *reflexed* r. *relaxed* ; p. 109. l. 18. for *hastening* r. *happening* ; p. 117. l. 28. for *Semicirculari* r. *Semicirculare* ; p. 119. l. 12. for *becomes* r. *become* ; p. 138. from *And therefore* in the 7th line to the end of that Paragraph, leave it out : p. 137. l. 7. for **above** r. *below* ; p. 168. l. 8. after *passage* add *at least* **but** *very little*.

www.ingramcontent.com/pod-product-compliance
Lightning Source LLC
Chambersburg PA
CBHW021526210326

41599CB00012B/1397